AF302051

Energie & Nachhaltigkeit

Herausgeber: Prof. Dr.-Ing. Kai Peter Birke

Band 12

Sustainable Production of Carbon Monoxide by Direct Current Gas Discharge

Von der Fakultät Informatik, Elektrotechnik und Informationstechnik
der Universität Stuttgart zur Erlangung der Würde eines
Doktor-Ingenieurs (Dr.-Ing.) genehmigte Abhandlung

Vorgelegt von
Stephan Renninger
geboren in Leonberg

Hauptberichter:	Prof. Dr.-Ing. K. P. Birke
Mitberichter:	Prof. Dr. Vera Meynen
Tag der Einreichung:	11.05.2022
Tag der mündlichen Prüfung:	26.10.2022

Institut für Photovoltaik der Universität Stuttgart

2023

Bibliografische Information der Deutschen Nationalbibliothek

Die Deutsche Nationalbibliothek verzeichnet diese Publikation in der Deutschen Nationalbibliografie; detaillierte bibliografische Daten sind im Internet über http://dnb.d-nb.de abrufbar.

1. Aufl. - Göttingen: Cuvillier, 2023
Zugl.: D93, Stuttgart, Univ., Diss., 2022

© CUVILLIER VERLAG, Göttingen 2023
Nonnenstieg 8, 37075 Göttingen
Telefon: 0551-54724-0
Telefax: 0551-54724-21
www.cuvillier.de

Alle Rechte vorbehalten. Ohne ausdrückliche Genehmigung des Verlages ist es nicht gestattet, das Buch oder Teile daraus auf fotomechanischem Weg (Fotokopie, Mikrokopie) zu vervielfältigen.
1. Auflage, 2023
Gedruckt auf umweltfreundlichem, säurefreiem Papier aus nachhaltiger Forstwirtschaft

ISBN 978-3-7369-7745-7
eISBN 978-3-7369-6745-8

Contents

Abstract

Carbon is the crucial element of organic chemistry and centerpiece in most produced chemicals and fuels. Its source is almost always fossil in nature. In light of climate change other paths to supply it must be developed. Carbon monoxide is a key molecule for carbon utilization, since forming it into hydrocarbons is an established process. It can be made using non-fossil feedstock and electricity. Next to electrolysis, biochemistry or electrically heated reactors, plasma technology is one main approach to this challenge. Plasma is used industrially, but rarely in process technology. Gliding arc plasma sources have the potential to change this, because they produce a 'warm' plasma - it allows for high chemical reactivity at low energy input and sufficient space-time yield. Low in this context means close to the reactions theoretical input or roughly $1...10\,J\,SCC^{-1}$. The first part of this work is dedicated to the construction and characterization for two successive iterations of gliding arc plasma sources and the electronics that are required to drive them. Voltage-current characteristics, discharge behaviour and geometry could be studied, while the driver circuits allowed for stable and efficient operation. Key characteristics for scaling of gliding arcs could be determined. In the second part, the plasma sources are used to facilitate the splitting of CO_2 into carbon monoxide and oxygen in a coaxial reactor that utilizes a packed bed of ceramics to quench the gas. This approach leads to high energy efficiency of up to 43% in the plasma while also converting 27% of the supplied CO_2. The results are compared to literature to give insight into the design criteria that allowed for this performance. Although impressive, at 43% energy efficiency half of the energy still ends up as heat. For this reason, two plasma processes are presented in the third part that utilize this heat: In the first, the calcination of a carbonates by plasma waste heat releases CO_2 that can subsequently be split in the plasma reactor. In the second process, heat is used to facilitate the gasification of organic solid wastes which are then reformed with steam or CO_2 in the plasma to obtain syngas. Both processes could be tested in preliminary experiments. Design criteria are worked out based on these experiments and the measured characteristics of the gliding arc plasma source. In the fourth part, all processes are evaluated economically using the calculated mass flow of a proposed simplified plant in a sensitivity analysis on the example of syngas production for methanol synthesis. The splitting of CO_2 is found to be a carbon sink when electrical energy is mostly renewable, but in all likelihood it is not economically viable in a large scale plant for syngas production in the near future. It needs to be paired with water electrolysis and arrives consistently at higher cost than a co-electrolysis that it is compared to. Plasma waste reforming could be determined to be an economically and environmentally viable alternative to the use of natural gas in steam reforming of methane.

Kurzfassung

Kohlenstoff ist das zentrale Element der organischen Chemie und damit auch in den meisten Chemikalien und Kraftstoffen enthalten. Er ist fast immer fossilen Ursprungs. Um den Klimawandel zu stoppen, müssen daher alternative Quellen erschlossen werden. Als Zwischenschritt dient dazu oft Kohlenmonoxid - ein Schlüsselmolekül vieler Syntheserouten, da es in etablierten Prozessen zu Kohlenwasserstoffen reagiert werden kann. Kohlenmonoxid kann aus nicht fossilen Quellen wie, Abfallstoffen oder Biomasse, mittels elektrischer Energie gewonnen werden. Neben Elektro-, und Biochemie und elektrisch beheizten Reaktoren ist die Plasmatechnologie ein vielversprechender Ansatz, um dies zu erreichen. Plasma wird oft industriell eingesetzt, jedoch selten in der Verfahrenstechnik. Gleitbogenplasmen sind dazu geeignet, da sie ein "warmes" Plasma erzeugen, welches eine hohe Reaktivität bei geringem Energieeinsatz und somit hohe volumetrische Ausbeuten erreicht. Gering bezeichnet in diesem Kontext einen Energieeinsatz nahe am theoretischen Idealwert oder ca. $1...10\,J\,SCC^{-1}$. Der erste Teil dieser Arbeit widmet sich der Entwicklung, dem Bau und der Charakterisierung zweier Iterationen von Gleitbogen-Plasmatrons, sowie der Treiberschaltung zu deren Versorgung. Strom-Spannungskennlinien, das Verhalten der Entladung und deren Geometrie konnten untersucht werden. Die Treiberschaltung ermöglichte die robuste und effiziente Versorgung des Plasmas. Die Schlüsseleigenschaften der verwendeten Gleitbogenentladung konnten ermittelt werden. Im zweiten Teil der Arbeit werden die Plasmareaktoren eingesetzt um darin CO_2 zu Kohlenmonoxid und Sauerstoff zu spalten. Das geschieht in einem koaxialen Reaktor, welcher ein Schüttbett nutzt, um das Gas schnell abzukühlen. Dieser Ansatz erweist sich als vielversprechend: Der Reaktor erreicht einen energetischen Wirkungsgrad von 43% bei einem Umsatz von 27%. Die Ergebnisse werden mit der Literatur verglichen, um zu ermitteln, welche Designentscheidungen zu den gemessenen Ergebnissen führen. Obwohl 43% Wirkungsgrad eine Verbesserung zum Stand der Technik darstellen, treten nach wie vor beträchtliche Verluste auf. Daher werden im dritten Teil der Arbeit zwei neue Plasmaprozesse betrachtet, die die Nutzung der Abwärme erlauben: Im ersten werden durch die Abwärme des Plasmas Carbonate, insbesondere Kalk gebrannt, wobei CO_2 freigesetzt wird. Dieses wird anschließend im Plasma gespalten. Im zweiten Prozess wird die Abwärme des Plasmas zu Vergasung von organischen Abfällen verwendet. Diese können anschließend im Plasma reformiert werden, um Synthesegas zu erzeugen. Beide Prozesse wurden in Versuchen qualitativ untersucht. Anhand der Ergebnisse und den zuvor gemessenen Eigenschaften der Gleitbogenentladung konnten Anforderungen an zukünftige Experimente erarbeitet werden. Im vierten Teil der Arbeit werden alle Prozesse von einem wirtschaftlichen Standpunkt betrachtet. Dazu wurden zunächst die Massenströme durch alle Komponenten von zu diesem Zweck vereinfachten Anlagen berechnet. Ziel ist eine Tagesleistung von 100 Tonnen Synthesegas für die Methanolherstellung. Diese werden techno-ökonomisch analysiert. Dabei zeigt sich, dass die CO_2-Spaltung durch Plasma eine negative CO_2-Bilanz aufweist, sofern erneuerbare Energien zu deren Versorgung eingesetzt werden. Sie ist jedoch vermutlich in der nahen Zukunft nicht wirtschaftlich konkurrenzfähig. Das wird dadurch verstärkt, dass eine Wasserelektrolyse zugeschaltet werden muss, um den für Synthesegas benötigten Wasserstoff zu erzeugen. Im Gegensatz dazu erweist sich die plasmabasierte Reformierung von Abfall bereits zu heutigen Marktbedingungen als konkurrenzfähig zur etablierten Reformierung von Erdgas.

Chapter 1

Introduction

The mitigation of climate change due to greenhouse gas emissions as well as the protection of the worlds biosphere are among the major challenges of this time. The manufacturing of chemicals is responsible for 6% of global direct CO_2 emissions, large parts of which could be avoided by the use of alternative feedstock [1]. Additionally, the transport sector emits 21% of CO_2 globally [2]. Both are supplied by the petrochemical sector, which accounts for 7% of global CO_2 emissions and 10% of energy uses [3]. It is not easy to exactly define how these sectors overlap, but simply adding the numbers leads to the conclusion that the industrial manufacturing and use of hydrocarbons as fuel in the transport sector and in production of chemicals contribute 34% of global CO_2 emissions. Unsurprisingly, carbon emissions appear when burning and manufacturing hydrocarbons, which cannot be replaced in either sector: As a fuel in mobile applications hydrocarbons feature the best combination of storability, energy density and non-toxicity. In chemicals, carbon is the very essence of organic chemistry. Without it, there wouldn't be lacquers, solvents, surfactants, or polymers to name just a few. For this reason, the chemical industry objects to the term 'decarbonisation' and instead refers to 'defossilisation'.

Fossil fuels contain the carbon and hydrogen that are used to synthesise hydrocarbons. One might argue that large parts of this carbon end up in products as far as the chemical industry is concerned. But a vast share of these products are not recycled, so carbon still ultimately ends up in the atmosphere. However, the carbon budget humanity has left to emit to keep global warming below 1.5°C is nearing its end [4]. The intergovernmental panel on climate change (IPCC) concludes that the most likely scenario for reaching this 1.5°C goal must reach zero effective emissions by 2050, which is illustrated in fig. 1.1. It must be concluded, that no new carbon should be digged up in whatever form, and that the industry must shift towards a true circular economy: Alternative carbon sources must be made available, all while not relying on fossil fuels as energy source.

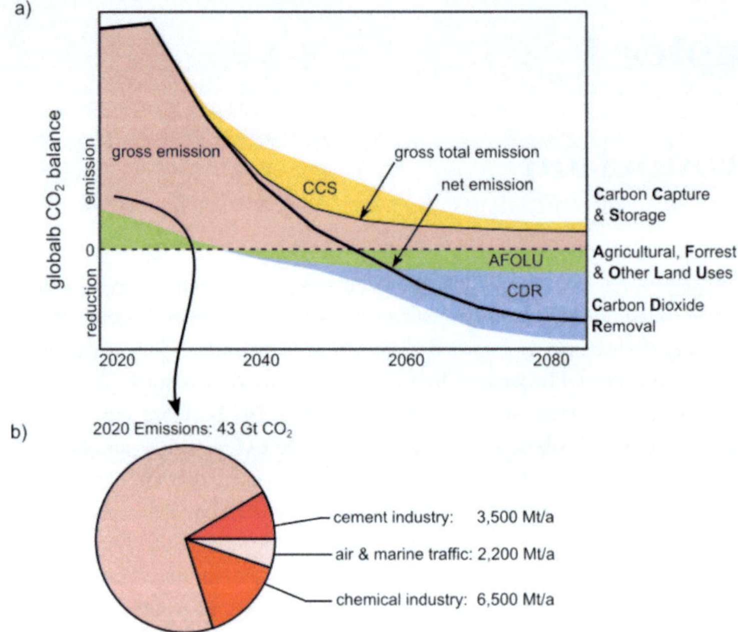

Figure 1.1: a) CO_2 emission goals including carbon dioxide capture (CCS) and removal (CDR) technology to achieve the 1.5°-goal, according to the 2017 IPCC report [5]. CO_2 emissions by sector are shown in b), highlighting the sectors that are addressed in this thesis and their share of global emissions.

Organic sources come to mind first; bio polymers are currently on the rise after all [6, 7]. They cannot compensate for fossil feedstock on their own. Plant feedstock rivals with food production for acreage and water consumption. Land use and eutrophication by agriculture are growing issues [8]. Add to that the fact, that the world produces 2 billion tons of sugar cane annually, the most farmed crop, but in the same time consumes 4 billion tons of oil [9, 10].

Carbon dioxide is gaining attention as a carbon source. Its emission is the main reason for man made climate change and the gas has become a recognized scourge of humankind on its own. Furthermore, the emission of carbon dioxide is expected to become prohibitively expensive in the near future thanks to mandatory emission certificates, at least in Europe. CO_2 of course is very energy intensive to utilize. After all, it is mainly emitted in energy production, and this would have to be reversed.

Anthropogenic carbon sources are a promising alternative that could offer a compromise between moderate energy requirements and availability. In fact, a lot of waste products created by modern civilization are carbon and energy rich, but some remain untapped until this day: Sewage sludge, scraps, residential and industrial waste end up in landfills or pollute the oceans, river systems and landscapes.
Biogas can be made from sewage sludge, plastics are sometimes recycled or down cycled, but the vast majority of these resources is only used as a low grade energy source, while the contained carbon still ends up in the atmosphere. The material use is still an exception. This is remarkable, considering the energy contained especially in waste plastics: The energy content of dry waste plastics is higher than that of crude oil [11].
After establishing a source of carbon, it must be converted into a useful form. Research is spread across multiple fields to develop technologies that are capable of doing this while using renewable energy sources: Bioreactors use microorganisms which in turn produce enzymes to convert CO_2 directly into useful products [12]. Photochemical processes use light to activate molecules. Electrolyzers can break CO_2 molecules to form more energetic ones like formic acid in a single reactor [13]. Conventional thermochemical processes may be driven by electrical heaters instead of gas fired boilers [13]. Plasma technology can be applied to drive endothermal reactions like the splitting of CO_2 in gas phase reactors [14]. The latter is the topic of this thesis.

1.1 Scope of this Work

The aim of this thesis is to demonstrate sustainable production of carbon monoxide (CO) by plasma. In a broader sense, the goal is to find an economic and ecologic path for carbon utilization. Carbon monoxide is only one out of a range of products that can be made from carbon dioxide, albeit the only one without using a source of hydrogen. However, it is a stable gas molecule and usually the product of all high temperature plasma processes; albeit not the easiest to work with chemical intermediate. Fig. 1.2 shows various intermediate chemicals with one carbon atom (C1-molecules) alongside Acetylene. Generally, the energy content is higher, if the oxidation state of carbon is lower; ranging from CO_2 at +IV to methane with -IV. Hot plasma can produce selectively only CO from CO_2 or Acetlyene from a hydrocarbon source [13, 15]. The other shown compounds require intermediate steps or cannot be made with high selectivity in plasma at all.

Mixing carbon monoxide with hydrogen gives syngas, from which a vast array of chemicals can be manufactured. Plasma based carbon dioxide splitting can be realized with a wide array of plasma technologies [14]. The initial goal of this thesis was the study and optimization of this process using a plasma process. After experimenting with dielectric barrier discharges (DBDs) for nearly two years, gliding arcs (GAs) were picked as more promising technology after an initial experiment surpassed most previously reported performance data for DBDs. Chapter 4 covers the characterization of two GA plasma sources that were constructed. This includes the reactors and electronics to drive them. They are used for the splitting of CO_2 in chapter 5. The results obtained are enticing. However, working on this topic more than three years overall also raised awareness of the problems that plasma-based CO_2 splitting inherently brings: Alongside technical challenges, the process is very energy intensive. For this reason, two alternative processes that build on the gliding arc plasma technology were devised. They are proposed and briefly discussed. The scope of the first process is the production of sustainable cement and carbon monoxide from calcium carbonate, while the second aims at production of syngas from waste plastics. Preliminary experiments and design consideration for future experiments are discussed in chapter 6. The proposed processes are supposed to drastically reduce energy requirements and deliver carbon feedstock at lower cost than the splitting of CO_2. Before large scale adaption, a process must proof that it is economically viable. Costs of plant, feedstock and energy have to be balanced. This is not only limited to monetary costs, but must also include direct and indirect CO_2 emissions. The sustainability and economic feasibility of the processes when compared to state-of-the-art fossil processes is discussed in chapter 7. To better understand how the presented processes might fit in the future circular chemical industry, a look on the possible resource chains of circular economy is necessary.

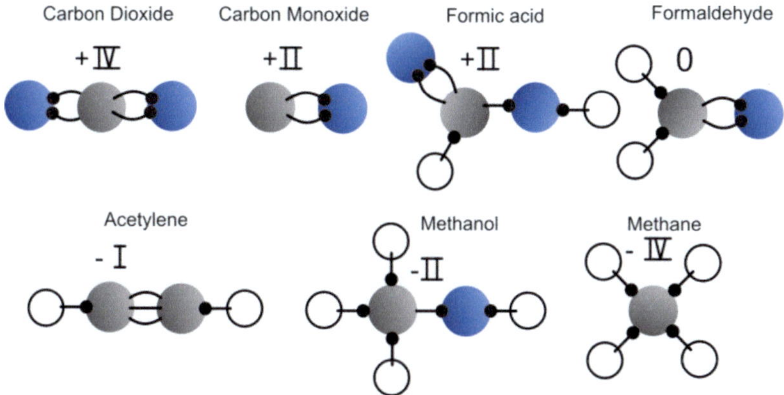

Figure 1.2: Carbon and its oxidation states in different C1-products (also Acetylene) [13, 16].

1.2 Resource Chains of the Circular Economy

Deriving chemicals from fossil feedstock offers a major advantage: The feedstock itself contains a lot of energy, which is emphasised by the fact that carbon is mostly present in oxidation state +IV. Most synthesis performed are either exothermal or only slightly endothermal. For this reason, not a lot of extra energy input is required when synthesizing desired products. This is explained in the following on the example of benzene [17], which is a precursor to nylon, phenol, polystyrenes, polycarbonates and expoxys – a large share of common polymers. Benzene can be obtained from catalytic reforming of hexane, which is the most common process used to date. A quick look at the very simplified reaction equation shows, that the energy requirements are low:

$$C_6H_{12} \longrightarrow C_6H_6 + 3H_2 \; \| \Delta H_0^R = +126 kJmol^{-1} \tag{1.1}$$

Hexane in turn can be destilled from crude oil at low cost. If benzene were to be made from carbon dioxide and water instead, the required energy is given by its enthalpy of combustion: A staggering 3.3 MJ per mol:

$$6CO_2 + 3H_2O \longrightarrow C_6H_6 + 7.5O_2 \; \| \Delta H_0^R = +3368 kJmol^{-1} \tag{1.2}$$

Processes for the production of benzene from biological feedstock like lignin are a topic of current research, however no industrial plant exists yet [18]. The energy requirement is between the fossil- and carbon dioxide based approach. An electrical process is theoretically possible via the intermediate product ethene – the final synthesis of benzene from ethene is well established [18]. This state is representative of most chemicals that are typically derived from fossil feedstock. The high energy requirement of carbon dioxide utilization methods makes the demand for energy efficient processes all the more apparent.

In contrast, not all chemicals are traditionally derived from fossil fuels. Methods like electrolysis already are well established industrial processes. An example for this is the Chloralkali process, which is utilized widely to produce most of the sodium hydroxide and chlorine used industrially today [19]. Along with the production of aluminum, the Chloralkali process is the most well established electrochemical process in the chemical industry. In contrast to petrochemistry, the energy input is almost exclusively electrical. Thus, these processes have the potential to become green due to the shift to renewable electric energy sources.

The goal of the chemical industry must be to achieve similar processes for organic chemicals. Electrolysis has been the only employed way to introduce electrical energy into chemistry for a long time, except for some niche applications. The developments in the fields of material science and electronics shifted focus towards more fields of research recently: Electrically heated high temperature catalytic reactors, photo-catalysis and plasma catalysis. Together, these approaches should be able to offer a suite of renewable precursor chemicals and intermediates that can be used to produce all products desired. The production chain to synthesize these products must be adapted to the new precursors. New synthesis routes become necessary, different intermediates will serve as stepping stones [13]. Economic viability can be increased by valorization of side products. New separation technologies are under development to extract products and side products.

1.3   Carbon Utilization Technologies

Carbon presents some unique challenges, due to its variety of stable oxidation states from -IV to +IV. This is especially challenging in electrolysis and leads to the requirement for highly tuned catalysts to obtain reasonable selectivity [20]. This is unlike water electrolysis, which is an established process: Hydrogen only has two occurring oxidation states, one of which it is already in as feedstock. Furthermore hydrogen as well as oxygen are always gaseous as a product. The equilibrium electrode voltage for water electrolysis is high at $1.23\,V$, but as long as the electrodes remain stable only the desired products are created. Carbon dioxide, on the other hand, can result in solid products like carbon black. The equilibrium voltage required for splitting carbon dioxide to monoxide is even higher at $1.47\,V$. For this reason, it is not ideal to utilize carbon dioxide via an electrolyzer. On the other hand, electrolysis can produce a variety of products like methanol or formic acid [13, 20], while plasma based processes usually result in carbon monoxide. A variety of products might be achieved in post plasma catalytic processes, but these did not reach significant success yet [21]. Another drawback of plasma based utilization is that produced oxygen is not automatically separated as it is in electrolysis.

1.4　Plasma Technology in the Chemical Industry

Plasma technology is in use in the chemical industry since the early 1900s. The first commercial application of plasma next to arc lamps were arc furnaces. Siemens obtained a patent for the generation of electric arcs using electrodes that are in principle copper pipes through which a coolant is pumped; other designs were also common [22, 23]. These were first used to produce calcium carbide from calcium oxide. For a long time, this was the primary intermediate in production of acetylene. Acetylene used to be the main source of carbon in organic chemistry and its production from coal hence an important industrial process. This was true especially in Europe. German chemist Walter Reppe was in large parts responsible for the invention of processes that allowed the production of vinyl, carboxylic acids, ethyline compounds and cyclic compounds like benzene [24, 25]. As late as 1980, direct synthesis of acetylene from coal in arc furnaces was implemented in pilot scale plants in West Germany via the Hüls-Process [26]. However, acetylene lost its importance when oil surpassed coal as the cheapest and most utilized feedstock after the second world war. Another example of electrical arcs in the chemical industry is the Birkeland-Eyde process [27], in which air is passed through an arc discharge. The resulting gas contains nitrogen oxides which can be converted to nitric acid in a packed column absorption tower. Birkeland and Eyde combined multiple reactors to multi-megawatt scale in Norway next to a hydroelectric power station to build what would be by today's standards a very sustainable chemical factory. However, the process was also supplanted by the Bosch and Ostwald processes through the prevalence of cheap natural gas and oil. Both acetylene and nitrogen oxides were manufactured using thermal electrical arcs at very high temperature. Non-thermal plasma in the form of corona discharges, sparks or dielectric barrier discharges saw its first use in the synthesis of ozone, where it is still in use on the multi-kW scale today [28, 29].

Within the 20th century, the production of chemicals by plasma became a niche application: Plasma was far more frequently used in the semiconductor industry and material sciences. However, the processing of semiconductor materials uses plasma generated at very low pressures, less than 1% of atmospheric pressure. The generation of plasma at ambient pressure holds different challenges. To enable chemical processes by plasma, ambient pressure operation is essential for two reasons: Firstly, compressors are expensive and require additional energy. Secondly, the volume of a low pressure plasma reactor would need to be very large in comparison to the mass of chemicals flowing through it [13]. The most prevalent ambient pressure plasma technology today are arc furnaces, which account for roughly a quarter of steel production. Arc furnaces produce a very thin and extremely hot plasma that is used to radiatively transfer heat into the material below. The small volume and extreme energy input make this type of plasma unsuitable for large scale chemical applications, not to mention the fact that the electrodes used degrade rapidly. New methods of plasma generation like inductively coupled and microwave sustained plasma emerged throughout the 20th century [30]. They were in large parts products of the invention of electron tubes and later semiconductor power electronics. These allow the generation of plasma with finely tunable properties. This is a stark contrast to previous arc devices, in which a ballast resistor or inductor often was the pinnacle of control electronics. Computational methods allowed for a deeper understanding of the processes taking place in plasma chemistry [31]. Thus, plasma turned from a way to provide heat to a reaction into a versatile method to introduce energy into a molecule by electron impact. The goal now is the generation of large volume plasma with defined properties, which offers the potential for more efficient chemical processes. Non-thermal plasma has gained attention for its catalytic properties, which can be obtained at moderate energy input. While plasma technology has made significant advances, so has thermo-, bio-, photo- and electrochemistry. Hence the list for chemical processes in which plasma is used on large scale in the chemical industry is no longer than it was decades ago. This might possibly change soon with the overdue electrification of the chemical industry.

Chapter 2

Theoretical Background

To ultilize plasma in chemical processes, background knowledge in physics, chemistry and electrical engineering is required; this chapter aims to first define the characteristics of plasma and then give the required background information in each of the three subjects in this order.

2.1 Plasma

A plasma is formed by any gas that contains enough mobile charge carriers to significantly alter its properties. This directly relates to plasma's first observable property: It is electrically conductive. Naturally, charge neutrality requires any macroscopic object to have a net zero charge: The same number of electrons and protons is contained in any volume [32]. This is true in any medium on any length scale exceeding the Debye length, which is less than 1mm in atmospheric pressure plasma. When both are abundant at low temperature, they tend to combine to form stable gas molecules. Thus, no charge carriers are available for charge transport and gas is non-conductive. The transition from a non-conductive gas to a conductive plasma is explained in section 2.1.1. The following section 2.1.2 gives an overview of the particles that can be found in a plasma and their different excited states. To utilized plasma in chemical processes, operation at ambient pressures is favourable. This is contrary to what is widely used and understood in practical applications such as neon lights. The additional effects that need to be considered then are discussed in section 2.1.4. Since magnetic fields were used to address some of the problems that arise from ambient pressure operation, the effect of magnetic fields on plasma are covered in section 2.1.5. Lastly, a few measurement techniques that were employed to characterize plasma in the scope of this thesis are discussed in section 2.1.6.

2.1.1 Formation

Free charge carriers can be created only when enough energy is put into a gas to liberate them from their bonds. This energy is supplied in one of multiple ways [33]:

- electric fields

- magnetic fields

- ionizing radiation such as UV or X-rays

- high temperature

- microwave or radio frequency radiation

The abundance of charge carriers in a plasma is defining for its behaviour. The relative concentration of charge carriers is given as degree of ionization α [32]:

$$\alpha = \frac{n_i}{n_i + n_n} \tag{2.1}$$

where n_i is the number of ionized particles and n_n the number of neutral particles.

Cosmic rays and background radiation always generate a very small number of charge carriers in the form of free electrons and ions. Their number is so small, that no notable change in properties occurs. A cubic centimeter of air at ambient pressure contains roughly 10^{19} molecules. Only $10^3...10^4$ of them are ionized due to background radiation, corresponding to a degree of ionization of 10^{-15} [34]. To observe the properties of plasma, a higher degree of ionization is necessary. Ionization by electric and magnetic fields is usually realized in technical applications. The background electrons are accelerated by an applied field. In collisions between electrons and neutral particles, energy is transferred from the electrons to the gas molecules. If this energy exceeds the ionization energy, ionization occurs. This increases the charge carrier density and thus the degree of ionization. A value of $\alpha = 10^{-4}$ is sufficient to generate other typical properties of plasma: Light emission, high chemical reactivity, sensitivity to magnetic fields and conductivity [35].

However, when no external power source is present the plasma quickly looses its properties due to the recombination of charge carriers. During recombination, the previously expended ionization energy is emitted as light. A plasma is called self sustaining, when more or an equal number of charge carriers are created than are lost due to recombination [36]. This distinction is especially important in setups using conducting electrodes: In a self-sustained discharge, the number of electrons liberated by the plasma's action from the cathode surface must equal the number of electrons that arrive at the anode. A non-self-sustaining plasma can be ignited and run continuously only when a source of free electrons is present [37], i.e. a heated cathode that can thermally emit electrons.

2.1.2 Excited Particles

Plasma must contain electrons and ions, while neutral particles often make up the majority by number. Energy must be present to frequently create these charge carriers, because they constantly recombine. The energy that is present in the plasma distributes itself among these particles, where it can take on various forms. Particles that absorb this energy are referred to as excited, because they have an internal state that contains this energy. Excitation is categorized into thermal and electronic. What is referred to as thermal energy on the macroscopic level is just kinetic energy of particles on the microscopic level. It can be present in three different ways: Translation (movement, momentum), vibration and rotation of particles, which is illustrated in fig. 2.1 for the example of a CO_2 molecule [38].

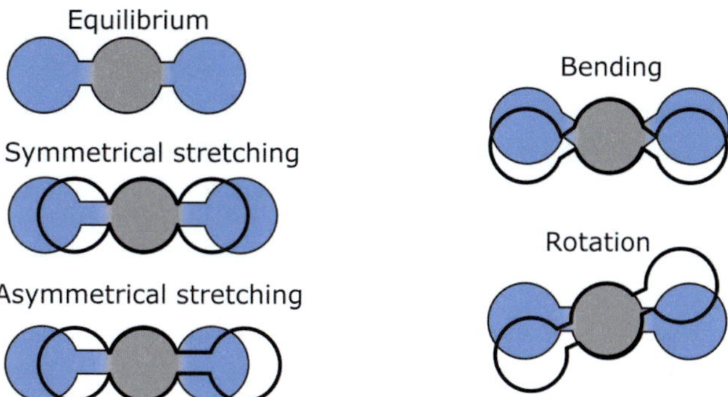

Figure 2.1: Vibration and rotation modes of the CO_2 molecule.

These three states are not equally excited at all temperatures, because vibrational and rotational excitation are noticeably quantized, i.e. energy can only be transferred into these states in fixed steps. The phrase 'noticable' is adequate in this context, since momentum also is quantized, but the energy levels are so numerous that it appears continuous. Rotational energy states lie further apart, as shown in fig. 2.2. Even further apart are vibrational energy levels. Combinations of states are possible [39].

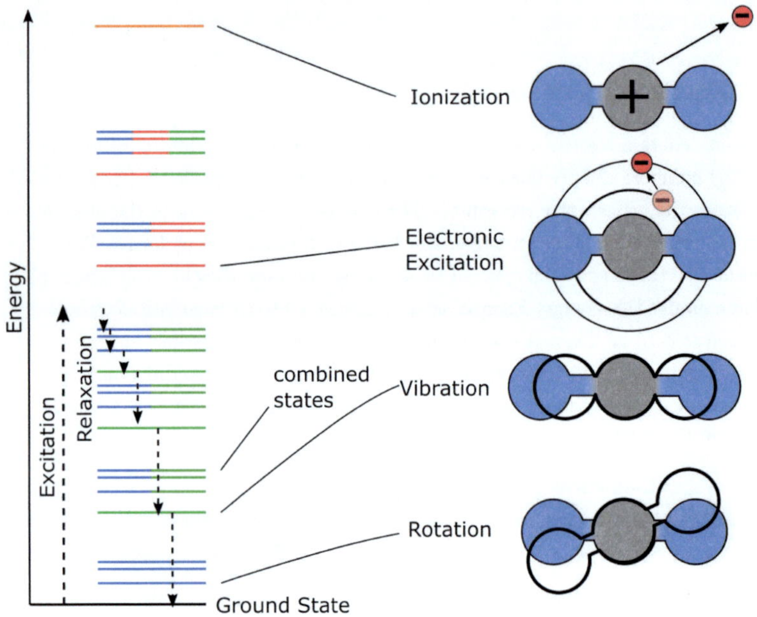

Figure 2.2: Schematic of modes of excitation in a CO_2 molecule and corresponding excitation energies [39]. Combined states are illustrated as multi coloured lines.

Both, rotational and vibrational energy states, are not excited at low temperatures at all; all energy is distributed in the translational energy states. As temperature increases, rotational energy states can be excited; only at very high temperature, i.e. greater than 2000 K, the vibrational states are excited as well.
Thermal energy will distribute evenly among all excited states. For this reason,

the heat capacity of a gas changes, as more states become available, which is illustrated in fig. 2.3. Each state theoretically contributes $c_{p,i} = 1/2\,R$ with $R = 8,31\,J\,mol^{-1}\,K^{-1}$. The total heat capacity is the sum of all and should vary between $c_p = 3/2\,R$ at ambient temperature and in theory $c_p = 7/2\,R$ for CO_2 at high temperatures [40]. This approximation becomes worse at high temperatures, because a significant thermal dissociation of CO_2 starts to occurs.

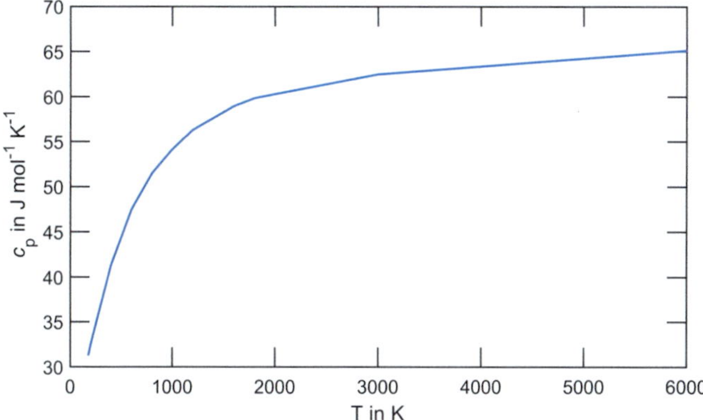

Figure 2.3: Relationship between heat capacity c_p of CO_2 and temperature T, data from [41].

Energy is transferred between these states by random collision between particles and interaction with photons. E.g. when a CO_2 molecule decays from vibrational states, the energy is released as mid to far infrared radiation [42]. Even more energy is required to electronically excite a molecule. In electronic excitation, an electron is transferred to a higher, but still bound, orbital. This can occur multiple times, until the electron is ultimately removed from the molecule entirely, resulting in ionization. Ionization is thus the highest possible energy state. Electronic excitation can only occur between orbitals and is thus also quantized.

Each of these states could be described with its own temperature. What is generally called plasma is a clutter of excited particles. However, as mentioned before, energy does distribute evenly among all available states given enough time. When it is evenly distributed the plasma is called 'thermal'. In consequence, this means that a lot of thermal energy has to be introduced into a system to populate all low energy states, before energy is introduced into electronic excitation and ionization. This is not ideal for the applications pursued by plasma catalysis, where electronic excitation should be performed with as little energy as possible.

It is possible to introduce energy into higher energy states directly [43]. This can be done by either light, by impact from high energy electrons or by collision between excited molecules. In the example of CO_2 splitting, energy can be introduced directly into the symmetrical stretching mode of the molecules [43]. Vibrational excitation is a promising path for CO_2 splitting, since it requires low activation energy compared to electronic excitation modes, which is shown in fig. 2.9. Ideally, all energy would be introduced into it. This leads to different temperatures for different states. When higher energy states are populated before the lower ones, the plasma is referred to as 'non-thermal'. Ionization and chemical activity can be achieved, while the neutral gas particles remain at low thermal energy, even as cold as ambient temperature. Since no energy is 'wasted' on kinetic and rotational energy states, non-thermal plasmas are often utilized in plasma catalysis. Sadly, non-thermal distributions of energy states degrade rapidly. The half-life of a vibrationally excited molecule depends on the density of the gas, but it can be in the sub-microsecond range [43]. This process in which energy is shifted from higher to lower energy states is called 'relaxation'.

2.1.3   Various Discharges & their Electrical Properties

In the previous section it has already been established, that a plasma needs to be continuously supplied with power to persist, and that this power can be introduced using electric fields. This is a convenient and easy method and also the one utilized in the scope of this dissertation. To properly introduce electrical energy into the plasma, the electrical characteristics are relevant.

Ohm's law can be written in a general form to calculate the current density J,

$$J = \sigma E, \tag{2.2}$$

where σ is the conductivity and E is the electric field. Conductivity σ in turn depends on charge carrier density n and mobility μ_e

$$\sigma = q_e\, n\, \mu_e. \tag{2.3}$$

where q_e is the charge of the electron. Electrons achieve a mobility that is magnitudes larger than that of ions, so these can be ignored. While this formula might seem simple, predicting the conductivity of a plasma is not simple at all. The number density of charge carriers depends on the overall number density of particles n and the degree of ionization. The number density n is inversely proportional to temperature due to the ideal gas law. The degree of ionization changes quickly when energy is introduced into the plasma.

Mobility is even harder to estimate. It also depends on the momentum transfer collision frequency, which states how often a charge carrier collides with another particle. This is also dependent on number density of particles. All in all, the behaviour is complex and highly nonlinear. It must be concluded, that experiments are more useful to characterize plasmas than models and calculations - especially for engineers.

The typical voltage and current characteristics of a discharge are shown in fig. 2.7, although the exact characteristics strongly depend on electrode material, pressure and working gas. It can be seen, that the discharge can be operated in different regimes. At low voltage - or conversely electric field - no ionization can take place initially. Background photons can liberate electrons from molecules. Positive ions are formed, that move towards the negative electrode. Electrons are accelerated by the electric field and collide with more neutral gas molecules. This liberates more electrons, an effect referred to as avalanche ionization. Electrons can excite gas molecules on impact. When these gas molecules relaxate, they release a high energy photon. If this photon hits the negative electrode, more electrons can be emitted by photo ionization. This process sustains the discharge.

Corona discharges appear around one electrode; the given explanation is for negative corona discharges which form around the cathode, but positive discharges are also possible with slightly different mechanisms. The avalanche effect is limited to the space around the electrode, where the electric field is stronger due to the curvature of the electrode. This region is referred to as ionization region [44].

Figure 2.4 a) summarizes the mechanisms that lead to a corona discharge. Ionization by background photons initiates the discharge [45]. Further ionization is spatially confined around the negative electrode, because the field further form the electrode is not strong enough to ionize more gas molecules [46]. Since no more ionization processes take place further from the electrode, electron attachment is the dominant process. It emits a faint homogeneous glow. Negative ions are created, that travel slowly to the positive electrode. This area of the discharge is hence dark, as shown in fig. 2.4 b). It is referred to as drift region [44]. Discharges of this type are also called dark discharges. Electrons can be emitted from the electrode by multiple mechanisms: Photo ionization, impact ionization and field emission [46, 47]. Fig. 2.4 c) gives an approximate distribution for electric field E, Voltage V, positive and negative charge carrier density $n+$ and $n-$ and current densities $j+$ and $j-$ in a corona discharge.

The corona discharge belongs to the townsend regime, where the electric field is uniform. This uniformity is only possible at low current however, i.e. the nanoampere range: Ions move much slower than electrons in a plasma, and they do so in opposite direction [36]. A positive space charge is thus formed near the negative electrode at high current density, which distorts the field. This space charge is more pronounced and moves closer to the electrode as the current increases, leading to a higher electric field there while it is lower in the bulk. This region of the discharge is called cathode fall and gives rise to a highly luminous layer close to the electrode [48]. The non-uniform field enables another mode of electron emission in addition to photo ionization: Ions are accelerated sufficiently in the cathode fall, that their impact may also liberate electrons from the cathode surface. The bulk region of the discharge with lower electric field is called positive column [48]. The mechanisms at play are illustrated in fig. 2.5 a), while the distribution of field strength and charge carrier densities is illustrated in fig. 2.5 c). Fig. 2.5 b) shows the typical light emission from a glow discharge. The negative glow may have a different colour, because in addition to emission from gas molecules, atoms from the electrode can be emitted into the plasma through ion impact, a process called sputtering. The thickness of the cathode fall layer is inversely proportional to the number density n. At high pressures, almost the entire volume of the discharge is occupied by the positive glow [49].

In conclusion, the electric field distribution in a glow discharge shows three major regions which persist throughout most operation conditions: In the cathode fall region, a high field accelerates ions towards the negative electrode, leading to more electron emission. In the positive column, the electric field as well as the power density and light emission is constant. In the anode layer, a moderate negative space charge forms, while the extent depends on electrode material and pressure. More regions can be identified in low pressure plasma, but they are not relevant in atmospheric pressure plasma as they shrink to be unidentifiably small.

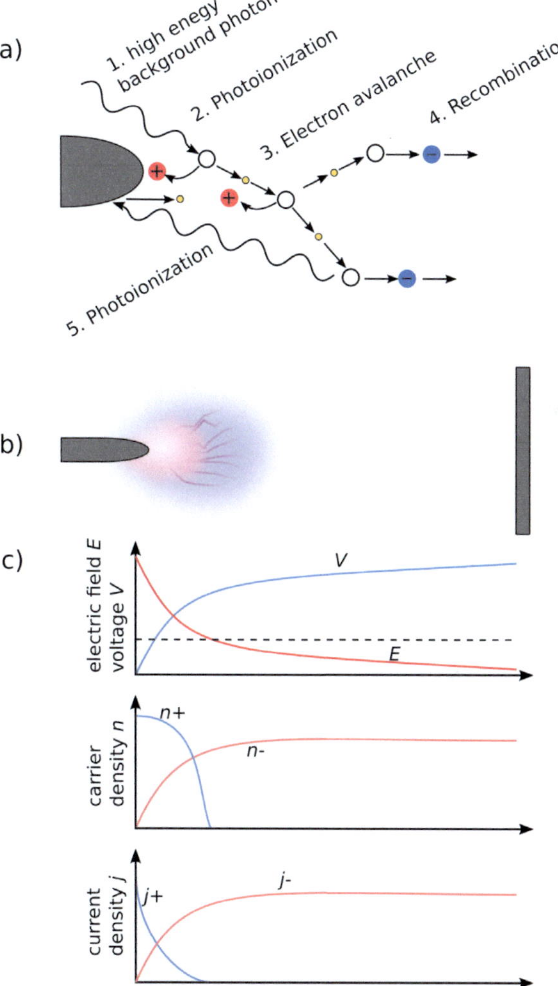

Figure 2.4: The mechanisms of creation and sustaining of a corona discharge are shown in a). Note, that various other mechanisms exist for secondary electron emission. In b), the visual appearance of a negative corona discharge are illustrated, while c) gives schematic representation of physical properties in the discharge [50].

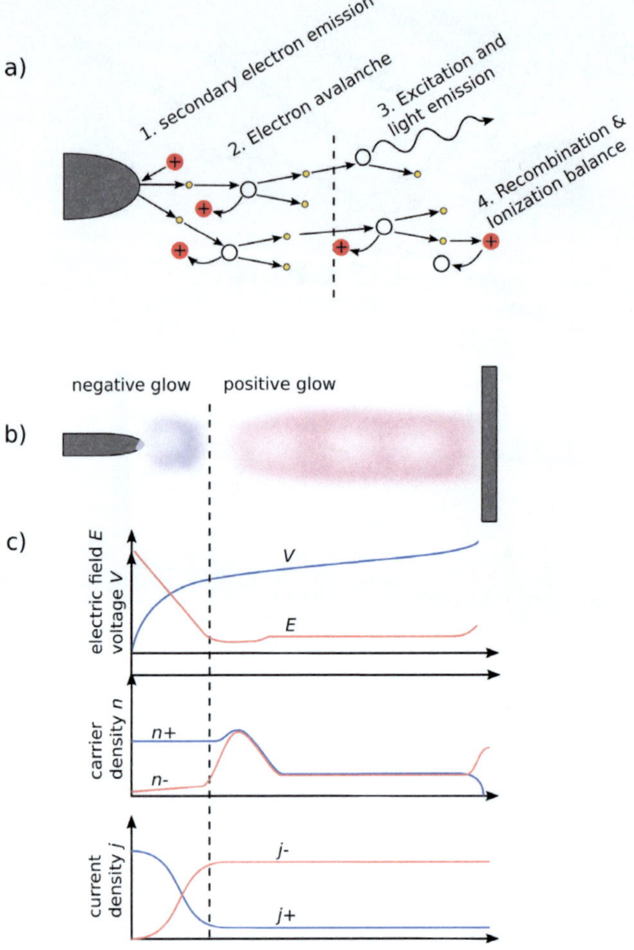

Figure 2.5: The mechanisms of creation and sustaining of a glow discharge are shown in a). In b), the visual appearance of a glow discharge are illustrated with a dotted line separating positive and negative region. c) gives schematic representation of physical properties in the discharge [36, 51].

Glow discharges persist over several orders of magnitude of current. At low current, the cathode zone covers only a small part of the negative electrode, which is called the cathode spot. When current is increased, this region grows while current density and voltage remain similar. This range is called the normal regime. Only after the whole cathode is covered, the voltage increases and a abnormal glow discharge is formed. This is characterized by an increase in electric field strength in the cathode fall. The increased cathode fall voltage helps to liberate more electrons from the cathode, which are necessary to sustain the current. The abnormal glow discharge drastically increases power density near the negative electrode [36].

This leads to heating of the electrode. When it reaches a sufficient temperature, the electron emission process and thus the type of plasma changes again: Thermal electrons can be emitted from the electrode at much greater current density than secondary electrons from collisions with ions [48]. This has several effects: Resistivity of the plasma drops drastically and with it electric field, because the degree of ionization increases and with it charge carrier concentration [36]. As the electric field provides less force to accelerate electrons, the mean electron energy decreases while the energy of neutral particles and ions increases due to the enormous power density. At the same time, the discharge contracts to a thin channel due to a positive feedback loop between gas heating and charge carrier production.The plasma transitions to a thermal energy distribution [48]. Fig. 2.6 a) shows the mechanisms at play, while b) illustrates the visual effects of an arc discharge. Fig. 2.6 c) gives an overview of the physical properties of an arc discharge.

All of these characteristic regimes and their electrical characteristics are summarized in fig. 2.7. The exact trend of voltage and power depends on the type of gas, the employed electronics and the electrode material and geometry. General observations can still be made: It is apparent, that only glow discharges and arc discharges can achieve meaningful power input, which is relevant in chemical applications. While thermal arcs occur alongside hot electrodes, glow discharges and non-thermal arcs utilize cold electrodes.

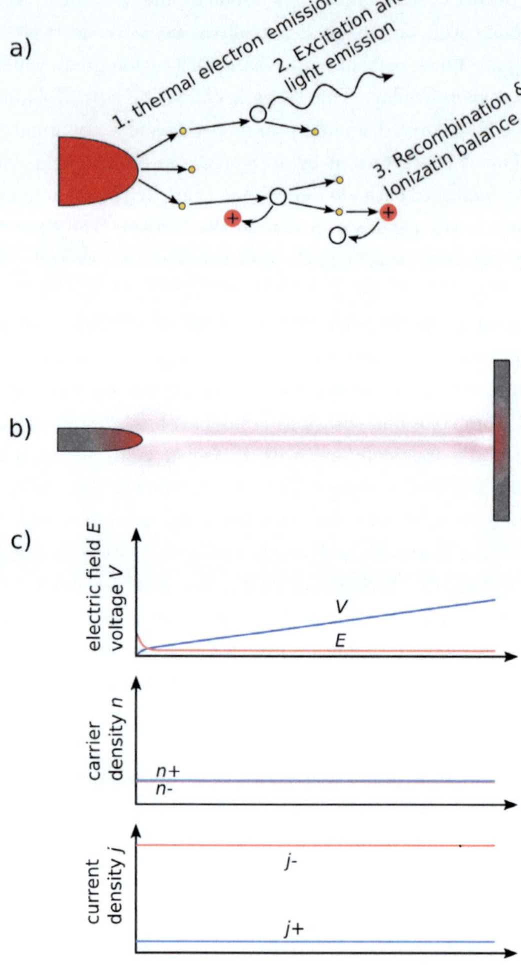

Figure 2.6: The mechanisms of creation and sustaining of an arc discharge are shown in a). In b), the visual appearance of an arc discharge are illustrated with a dotted line separating positive and negative region. c) gives schematic representation of physical properties in the discharge.

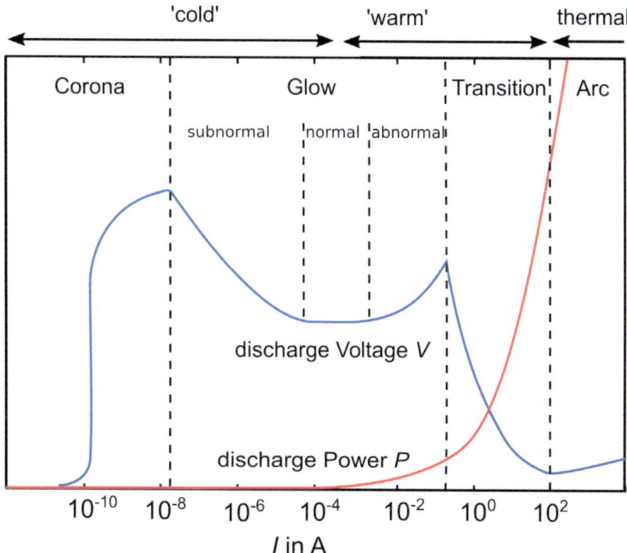

Figure 2.7: Schematic V/I curve of a typical gas discharge in pin to ring electrode configuration. Voltage and Power are given in arbitrary units [36, 37].

2.1.4 Plasma at Elevated Pressure

Large volume plasma has been used in material science industrially mainly at low pressures, e.g. for the treatment of silicon wavers. Elevated pressure changes the properties of plasma. One notable difference can be recognized before the plasma is even ignited: The breakdown field strength of the gas. It is described by Paschen's law [33]. While at near vacuum and milimeter-gaps a complicated behaviour is found, luckily the law can be simplified to calculate the breakdown voltage at atmospheric pressure to

$$V_B \sim p\,d \tag{2.4}$$

using pressure p and discharge gap d. This meas that the breakdown voltage is growing linearly in proportion to gap distance and pressure. The breakdown field strength for air at ambient pressure is roughly $E_B(air) = 3\,MV\,m^{-1}$, it is nearly the same for CO_2 [52]. However, it is lower for helium and other noble gases, since they have lower electron affinity. This means, they are less likely to bind free electrons, which delays the formation of a discharge. For this reason, noble gases can be used to ignite a plasma with less voltage [33]. This fact is not only used in neon sign lamps, but also in plasma generators. The properties of the electric field also have an influence on the breakdown voltage - if asymmetrical electrodes are used, polarity has an influence. Furthermore AC plasmas ignite at lower voltage than DC [53].

After ignition, the discharge behaves differently depending on the pressure as well. At sub-atmospheric pressures, a homogeneous glow can be observed over many orders of magnitude of applied current, filling the entire space available between the electrodes. On the contrary the discharge channel contract to a smaller radius in atmospheric pressure discharges. These millimeter sized channels are often referred to as filaments. Two main reasons can be named for this contraction [54, 55]: Firstly, the finite thermal conductivity of gases causes uneven heating. Since higher temperature leads to more ionization and thus higher conductivity, the effect is self-reinforcing. This behaviour is referred to as plasma instability. Low pressure plasmas with gas pressures in the range of millibars, feature significant diffusion due to the long mean free path of particles and the heat can spread more easily than in atmospheric pressure plasma.

Secondly, the pinch effect acts on all electrons that move collectively [48]. Moving charges create magnetic fields. The magnetic fields of two charge carriers moving in the same direction can be illustrated by the right hand rule (using two right hands, you'll have to ask a friend). The magnetic field interacts destructively between the charge carriers and constructively in the space around them. Since moving charge carriers tend to lower magnetic field regions, the electrons are pulled closer to each other [56]. This is especially noticeable at high ionization rates and with fast moving charge carriers, which can be found in high current plasma, i.e. several 100s to 1000s of amperes. Many properties of plasma depend on the ratio of electric field E and number density of particles n. Since it describes the behaviour of a plasma well in a variety of situations, the reduced electric field E/n is often used as a characteristic property. It is measured in Townsend, or Td. One Townsend corresponds to an electric field of $E = 2.5 \cdot 10^4 \, Vm^{-1}$ at atmospheric pressure, where $n = n_0 = 2.7 \cdot 10^{25} \, m^{-3}$. So $1\,Td = 10^{-21} \, Vm^{-2}$. E.g. at low reduced electric field, electrons typically acquire lower energy. Electron temperature corresponds to electron energy; it is usually given in $[T_e] = eV$, which can be converted into Kelvin by using the ratio of Boltzmann's constant and the elementary charge:

$$1\,eV \cong 11,605\,K \tag{2.5}$$

2.1.5 Magnetic Fields

All moving charged particles create and interact with magnetic fields. In the case of the previously described pinch effect, the plasma itself can create a magnetic field [48]. External magnetic fields can be used to change the paths of charge carriers within the plasma under certain conditions. Ideally, the Lorentz force acts on moving charge carriers at right angles to both their velocity and magnetic field [56]. This creates a movement known as ExB drift (ExB refers to the cross product of magnetic and electric field). If no collisions occur, the charge carriers thus evolve in a circular motion, while the centre of the circle, the 'guiding centre', moves in the direction at a right angle to both the electric field and magnetic field. The radius of the circle of motion is called Gyroradius or Larmor radius [56]. This motion is illustrated in fig. 2.8. In a coaxial electrode setup with an electric field that is radial, i.e. in a magnetron, electrons in the absence of collisions and with a sufficient magnetic field move in circles, never reaching the opposing electrode. This is not the case in ambient pressure plasma. The collisions lead to loss of momentum of charge carriers at random moments, which leads to disappearance of the circular motion. In effect, electrons move in a direction between the electric field vector and that of the magnetic field. The frequency of rotations is given by the mean free path, λ.

A rule of thumb can be given for when a significant ExB drift is expected: the gyration radius should be smaller than the mean free path [56]. This means, the electrons can travel a significant proportion of a revolution before being deflected by a collision. To get a feeling of the quantities involved: In the example of an atmospheric pressure plasma with a magnetic field of $B = 20\,mT$ and an electron temperature of $T_e = 5\,eV$ the gyration radius can be calculated according to [56] to be

$$r_G = \frac{M_e\,v_e}{q\,B} = \frac{10^{-30}kg\,10^6m\,s^{-1}}{10^{-19}\,C\,0.02\,T} = 300\,\mu m. \tag{2.6}$$

Here, M_e is the electron rest mass and v_e the velocity of an electron that can be calculated from its energy. q is the elementary charge. The mean free path on the other hand can be calculated usign a simplified equation from [57]:

$$\lambda = \frac{k_B\,T}{p\pi r_i^2} = \frac{1.4 \cdot 10^{-23}JK^{-1}2000\,K}{10^5 Pa\,\pi(10^{-10}m)^2} = 90\,\mu m. \tag{2.7}$$

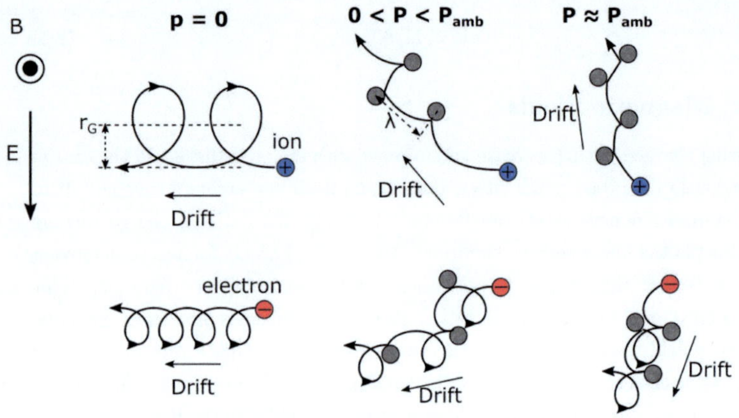

Figure 2.8: ExB drift of an electron and ion at different pressures. Mean free path λ decreases from left to right, while gyration radius remains constant. Note that the gyration radius r_G is larger for the ion. Adapted from [56].

Here, k_B is the Boltzman constant, T is the gas temperature, p is gas pressure and r_i is the ionic radius. All quantities are very crudely rounded. It can be seen, that the calculated gyration radius is larger than the mean free path, so no significant drift effect is expected [57].

Instead, a discharge channel at sufficient gas density and conversely pressure can be treated as a wire moving through a gas [58]. Hence, the Lorentz force F_L acting on each unit of length of the discharge channel can be calculated using

$$\frac{F_L}{d} = I \times B = I\,B \tag{2.8}$$

when the magnetic field is at a right angle to the movement of charge carriers. The discharge channel thus begins to move. This movement is oppose by air resistance, or drag, since the electrons loose large amounts of momentum to neutral gas particles. The drag force per length F_d can be calculated similarly to that of a wire as

$$\frac{F_d}{d} = \frac{1}{2} c_d\, \rho_{gas}\, v^2\, D, \tag{2.9}$$

with the discharge channels diameter D, drag coefficient c_d, and its velocity v [58].

2.1.6 Measurement Techniques

The character of an electrical discharge has a huge influence on the chemical processes that can be realized. Potential in the plasma can be measured as voltage relative to the electrodes. Ideally, a Langmuir Probe is used to measure the local voltage $V_{measured}$ at any place in the plasma. It consists of a wire loop tip placed on an insulating ceramic tube [59]. The voltage of the probe can be tuned relatively to one of the electrodes. A current flows from the probe into the plasma. By sweeping the voltage, a current waveform can be acquired. This waveform can be used to obtain information on electron density, potential and more. However, a simpler solution is used in this work for lack of a Langmuir probe: A straight wire probe with very high resistance can be coupled directly with an oscilloscope, forming a voltage divider. This way, the local potential in the plasma can be measured, albeit with less accuracy. The current flowing into the wire probe distorts the electric field in the plasma. A sheath layer is formed around the wire probe, leading to more inaccuracies. The method gives crude estimates, as long as the current through the probe is orders of magnitude lower than the current flowing through the plasma around the probe which is achieved by using a $10G\Omega$ resistor. In the scope of this thesis, a wire probe was used nonetheless, because no sophisticated equipment was available. By moving the probe along one axis x by a distance d, the electric field $E(x)$ in the plasma can be measured:

$$E(x) = \frac{\delta V_{measured}}{\delta d}.$$

(2.10)

The electric field can then be used to determine the properties of the plasma, e.g. to determine if the positive column has a constant electric field and if a cathode fall exists. Since sheath effects should remain approximately constant, they do not manifest in the measurement of electric field. To estimate power density within the plasma, knowledge of the current density is necessary. Under the assumption that current is homogeneously distributed throughout the discharge cross-section A, one can calculate current density J:

$$J = \frac{I_{discharge}}{A}.$$

(2.11)

The power density is calculated with

$$P_{vol} = J(x)\,E(x).$$

(2.12)

More sophisticated measurements are necessary to estimate the temperatures present in a plasma. As previously discussed, various species have different temperatures. They can be measured individually by in-operando Raman spectroscopy. A simpler technology is the use of light emission in the infrared and optical spectrum. This may rely on the addition of trace gases like hydrogen. For example, the relative brightness of different emissions, like in the case of hydrogen the Lyman alpha and beta emissions, can be used to estimate electron temperature. Sadly, none of these technologies were available during the experiments described in the following chapters.

2.2   Chemistry

The second range of topics that needs to be covered is chemistry. The carbon dioxide reduction reaction has complex underlying mechanisms, which are shortly covered in section 2.2.1. In section 2.2.2, the basic chemical quantities that are required to describe a chemical process and their derivation are covered. At high temperatures, i.e. several thousand degrees K, and in the presence of excited particles, chemical reactions rarely go to completion. For this reason, the underlying mechanisms that lead to the final concentration of chemicals in the exhaust gas of a plasma reactor are briefly explained in section 2.2.3.

2.2.1   The CO_2 Reduction Reaction

The reduction of carbon dioxide is described by the reaction equation:

$$CO_2 \longrightarrow CO + \frac{1}{2} O_2 \,|\, \Delta H_r^0 = +282.5\,kJ\,mol^{-1} = 2.93\,eV \qquad (2.13)$$

This equation describes the observable quantities of the reaction. The mechanism is however not well described by it. Different mechanisms exist for the splitting of CO_2 in plasma, an example reaction equation is given below for four examples [60]:

Recombination of ions with electrons:	$CO_2^+ + e^- \longrightarrow CO + O$
Dissociation after excitation:	$CO_2^* \longrightarrow CO + O$
Reaction with oxygen radical:	$CO_2 + O^* \longrightarrow CO + O_2$
Recombination of ions:	$O^- + CO_2^+ \longrightarrow CO + O_2$

Depending on the characteristics of the plasma that is used, different reactions account for most of the overall CO_2 conversion [43]. Of course, the energy needed is always higher than the standard enthalpy of reaction. For example, the dissociation of CO_2 after vibrational excitation produces atomic oxygen which is by itself an energetic state, and thus requires $\Delta H_{vib} = 5.5\,eV$. Thus, the energy efficiency for vibrational excitation induced splitting of CO_2 would not exceed the ratio of the two energies which is equal to 53% [61]. This is also illustrated in fig. 2.9. A full $2.57\,eV$ per molecule would thus go to waste. However, plasma systems have been reported to operate at higher energy efficiencies. This is due to the reactivity of atomic oxygen which is formed in the splitting reaction. The huge increase in energy requirement is partially offset by the followup reaction between atomic oxygen and CO_2, which is slightly endothermal and also yields CO. Simulations predict, that this boosts the upper limit of energy efficiency up to 86% [61]. The enodthermal reaction likely proceeds when CO_2 is in an already excited state.

However, the atomic oxygen can react with carbon monoxide or with oxygen as well, which results in adverse effects on efficiency. Since plasma creates many excited species and thus high reactivity, it is hard to pinpoint the exact chemical processes. In practice, Kozák et al. [60] simulated a whopping 205 contributing chemical reactions and thus found a set of 17 which can explain the splitting of CO_2 in dielectric barrier discharges approximately, at least under certain circumstances. They identified two major paths for the splitting reaction:
In the first case, high energy electrons transfer enough energy in a single collision to instantly dissociate a CO_2 molecule. This is likely to happen in plasma with high electron temperature, an electron needs at least an energy of $7.4\,eV$ to trigger this reaction. A highly reactive oxygen atom is produced. It is electronically excited, with one electron in its ^1D-orbital.

In the second case, the low vibrational states are excited by low energy electron impact. Collisions between excited CO_2 molecules lead to excitation of higher vibrational states. This process is referred to as ladder-climbing [43]. Ultimately, the dissociation takes place. In this case it can result in an oxygen atom with lower electronic excitation, with one electron in its ^3P-orbital. The overall energy that needs to be provided to the molecule is thus lower at $5.5\,eV$. Both processes are illustrated in fig. 2.9. In conclusion the energy efficiency that can be achieved in the CO_2 splitting reaction depends on the reaction mechanism. Multiple reactions are at play, which rely on various excited and intermediate species such as carbonate ions, atomic oxygen or ionized CO_2 molecules. Their concentrations in the plasma are subject to chemical equilibria, which will be discussed in section 2.2.3. At low reduced electric field, electrons typically acquire lower energy. This results in an effective transfer of energy into vibrational excitation states. Higher reduced electric field leads to electronic excitation and direct dissociation [43].

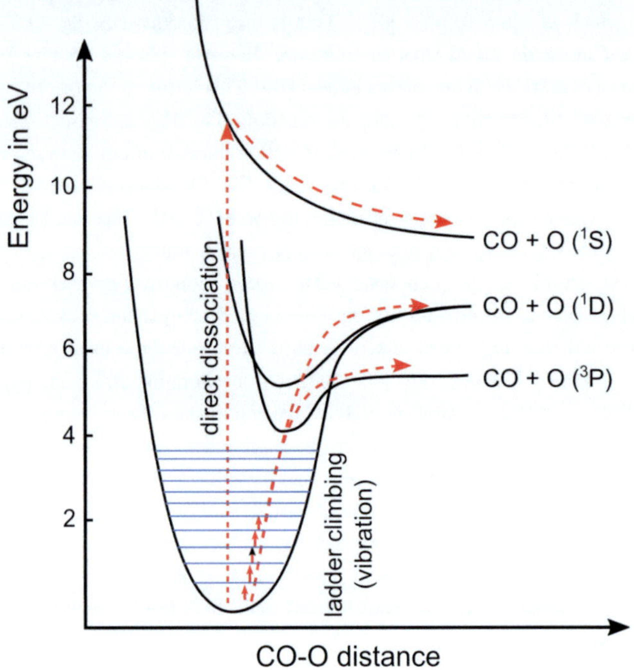

Figure 2.9: Schematic diagram of dissociation processes in a CO_2 molecule, illustrating CO-O distance as function of excitation energy relative to the molecules ground state for different dissociation processes. Adapted from [43].

2.2.2 Relevant Quantities and their Derivation

All particles produced from CO_2 in a plasma recombine and relax into stable or meta stable molecules, CO, CO_2, O_2 and O_3 in fractions of a second. Ideally, the only exhaust products of the CO_2 splitting reaction would be CO and O_2. Since backreactions and chemical equilibria, which are covered in the following section, do not allow for the conversion of all gas, the conversion rate X is used to measure the amount of CO_2 that was converted. A general equation for conversion of a reaction $A \longrightarrow B$ is

$$X = \frac{n_B(start) - n_B(end)}{n_B(start)} \tag{2.14}$$

where 'start' marks the beginning of the reaction and 'end' its end. Conversion in this work is calculated as

$$X = \frac{c_{CO}(end)}{c_{CO2}(end) + c_{CO}(end)}, \tag{2.15}$$

where c_{CO} and c_{CO2} are the concentrations of carbon monoxide and carbon dioxide in the exhaust gas of the plasma reactor. This is equivalent under the assumption that the starting material is pure CO_2 and that no carbon is lost in side reactions, as shown in appendix A. The quantities $c_{CO}(end)$ and $c_{CO2}(end)$ are easy to measure. Useful energy is lost during backreactions and converted to heat. An energy efficiency can be calculated, it is the ratio between enthalpy of the CO_2 splitting reaction H_r^0 and the expended energy W_{use}, which is electric energy in the case of plasma based splitting reactions. In the case of a direct current plasma with constant voltage V and current I, it can be calculated as

$$\eta = \frac{W_{use}}{W_{expended}} = \frac{\Delta H_r^0 \, n_{CO}}{V \, I \, t}, \tag{2.16}$$

where t is time in seconds. The amount of a substance n in mol is impractical to measure. Usually, the flow rate of the gas $\dot{V}_{CO2,in}$ is monitored. It is referred to in $SCCM$ or standard cubic centimeters per minute. Normalizing the reaction enthalpy to $J\,SCC^{-1}$ or Joules per standard cubic centimeters allows the calculation of useful energy per second; useful power P_{use}:

$$P_{use} = \Delta H_{r,vol}^0 \, \dot{V}_{CO,ex}, \tag{2.17}$$

where $\dot{V}_{CO}$ is the flowrate of carbon monoxide in the exhaust gas. This flowrate can be calculated using the conversion and overall input gas flow rate:

$$\dot{V}_{CO,ex} = \dot{V}_{CO2,in} \, X \tag{2.18}$$

The reaction enthalpy per unit volume $\Delta H_{r,vol}^0$ can be calculated using the molar volume of the gas V_M which in term is given by the ideal gas law, using standard conditions of pressure $p = 10^5 \, Pa$ and temperature $T = 273.15 \, K$:

$$pV = nRT \rightarrow \frac{V}{n} = V_M = \frac{RT}{p} = 22400\,SCC\,mol^{-1}. \tag{2.19}$$

$R = 8.31\,J\,mol^{-1}\,K^{-1}$ The reaction enthalpy per volume of the CO_2 splitting reaction then is

$$\Delta H^0_{r,vol} = \frac{\Delta H^0_r}{V_M} = \frac{282.5\,kJ\,mol^{-1}}{22400\,SCC\,mol^{-1}} = 12.6\,J\,mol^{-1}. \tag{2.20}$$

Using these equations allows the calculation of energy efficiency η of a plasma based CO_2 splitting reaction using only easily measurable quantities:

$$\eta = \frac{P_{use}}{P_{expended}} = \frac{\dot{V}_{in}\Delta H_{r,vol}X}{VI}. \tag{2.21}$$

Note, that $\dot{V}_{in}$ must be converted from SCCM to standard cubic centimeters per second with an additional factor for this equation to work.

Although the splitting of CO_2 is hard to achieve, it has favourable properties as well. The fact that almost exclusively carbon monoxide and oxygen are produced is not to be taken for granted. Often, more than one reaction product is possible. In these cases, not only conversion is important but also how much of the reactant is converted to the desired product and how much to a side product. The ratio between these quantities is described by the selectivity S. When a reaction can have two possible outcomes, such as

$$A \longrightarrow B$$
$$A \longrightarrow C \tag{2.22}$$

the selectivity is defined as the ratio of the amounts of B and C that are produced:

$$S(B) = \frac{n_B}{n_C + n_B}. \tag{2.23}$$

The selectivity for carbon monoxide in the CO_2 splitting reaction is close to one. It is slightly diminished by traces of carbon that can be formed. Yield describes the amount of product that is formed relative to the amount that could be formed ideally by the desired reaction (in this case, $A \longrightarrow B$). When selectivity is 1, yield Y is equal to conversion X. Otherwise, yield is the product of selectivity and conversion:

$$Y(B) = X(A)\,S(B). \tag{2.24}$$

2.2.3 Chemical Equilibrium and Plasma

A chemical reaction always progresses with a certain reaction rate v, which can have different units depending on context. It describes how much the number concentration of a reactant changes per unit volume and time. Reactions like the oxidation of carbon monoxide usually go to completeness under ambient conditions. For this reason it is sufficient to use simple reaction equations, such as

$$CO + \frac{1}{2}O_2 \longrightarrow CO_2. \tag{2.25}$$

The rate of this reaction is v_{back} equal to the change in concentration of carbon monoxide. While trying to split CO_2, the opposite reaction is desired. It's speed of reaction depends on the availability of high energy electrons and is referred to here as $v_{splitting}$. Concentration of carbon monoxide increases, as long as $v_{splitting} > v_{back}$. But reaction speed is not constant. It depends on the concentration of available reactants and the reaction rate constant k [62]:

$$v_{back} = k_{back}\, c_{CO}^x\, c_{O_2}^y$$
$$v_{splitting} = k_{splitting}\, c_{CO_2}^z \tag{2.26}$$

When both speeds of reaction are equal, the concentrations reach a steady state. This steady state confines the maximum conversion that can be achieved in the CO_2 splitting reaction, beyond which further energy input will not result in additional conversion. x,y, and z are the partial reaction order, their value reflects the reaction mechanism and interaction of the molecules. The reaction rate constant k is dependent on temperature, in the case of a plasma catalyzed reaction electron density, electron temperature and many more parameters influence it, too [63].

In a thermal plasma, where all components of the plasma have similar temperature, a temperature dependent equilibrium concentration $c_{CO,eq}$ can be calculated, which is shown in fig. 2.10 as illustrated by Bogaerts et al. [64].

This thermal equilibrium can be overcome by the use of non-thermal plasma. Put simply, the reaction rate constant of the back reaction k_{back} is kept low by the moderate gas temperatures, while high electron temperatures lead to a high reaction rate constant for the splitting reaction $k_{splitting}$. Several operation points achieved by various research groups are indicated in fig. 2.11 alongside the thermal efficiency limit previously calculated.

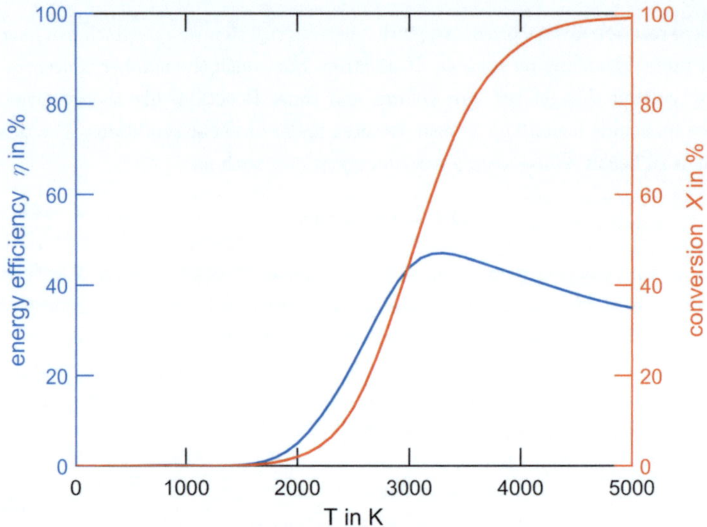

Figure 2.10: Thermal equilibrium conversion of CO_2 to carbon monoxide and oxygen at different temperatures, along with energy efficiency of the process, calculated assuming cooling to ambient temperature without recombination. Reproduced from [64].

Admittedly, there is not really a way around thermodynamics. A hypothetical full thermodynamic and reaction kinetic consideration including all species in the plasma would also lead to the observed results. It can be seen, that plasma based splitting of CO_2 can exceed thermal energy efficiency at low conversion. A likely reason is, that at low energy input the gas remains cold. Thus, the rate of the backreaction is also low. However, high conversion can only be achieved at high specific energy input. This also leads to heating of the gas. For this reason, the temperature of the gas should be brought back down to close to ambient temperature as fast as possible after the gas leaves the reaction zone. This reduces the amount of carbon monoxide lost to the back reaction. The process is called quenching and is especially important when working with thermal plasma. Heat removed from the gas during quenching may be reused for other chemical processes.

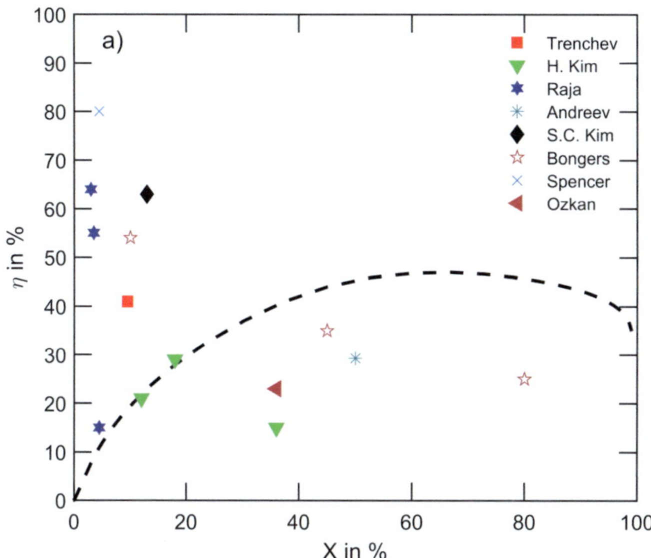

Figure 2.11: Literature operation points of plasma CO_2 splitting at ambient pressure (filled symbol) and low pressure (empty symbol) next to the thermal limit. Data from Trenchev et al. [65], H. Kim et al. [66], Raja et al. [67], Andreev et al. [68], S.C. Kim et al [69], Bongers et al. [70], Spencer et al. [71], and Ozkan et al. [72].

2.2.4 Chemical Reaction Systems

The previous section is concerned with a simplified case, in which only one reaction takes place. As soon as more species are added into the mix, multiple reactions can occur. The products are subject to thermal equilibria of all possible reactions [73]. The thermal equilibrium of product concentrations can be calculated in two ways. The first and most general approach is the minimization of Gibbs free energy in the gas mixture. The Gibbs energy of a mixture is the sum of the chemical potential of each species, which in turn depends on concentration, enthalpy of formation, standard entropy, pressure and temperature [73]. The second and perhaps easier approach was applied in this work. First, chemical reactions are found that relate all compounds present in educt and product gas are found. Then, the equilibrium constants $k(T)$ for each reaction depending on temperature are calculated. Each component of the product gas must appear in at least one reaction. Then, initial concentrations are set. The result is a equation system which should not be overdefined. It can be uniquely solved numerically to retrieve the concentration of each species in the product gas.

Two reactions influence thermal equilibrium in a reactor where steam and carbon monoxide are present: The Boudouard equilibrium and the water-gas-shift reaction. The Boudouard reaction is

$$C + CO_2 \longleftrightarrow 2CO \,. \tag{2.27}$$

The equilibrium constant $k_B(T)$ is dependent on temperature T in Kelvin. It can be approximated according to [74] by

$$k_B = \frac{c_{CO}^2}{c_{CO2}} = \frac{21574}{T} - 0.476ln(T) - 3.52 \cdot 10^{-4} T + \frac{98100}{T^2} + 25.66 \,. \tag{2.28}$$

The equilibrium constant for different temperatures and the resulting equilibrium conversion X is shown in fig. 2.12.

Comparing the equilibrium conversion shown in fig. 2.12 and the thermal CO_2 splitting equilibrium shown in fig. 2.10 illustrates, why usually none or very little carbon precipitation is expected in the CO_2 splitting reaction: Meaningful thermal conversion occurs above $2000\,K$, at which point the Boudouard equilibrium already reaches full conversion.

The water gas shift is a less temperature dependent reaction, that can convert carbon dioxide and hydrogen into water and carbon monoxide:

$$CO_2 + H_2 \longleftrightarrow H_2O + CO \tag{2.29}$$

At low temperature, the reaction favours the formation of hydrogen. At high temperature, carbon monoxide and water are favoured. The equilibrium constant of the water gas shift reaction (WGS) k_{WGS} can be approximated according to [75] by

$$k_{WGS} = \frac{c_{H2}\, c_{CO2}}{c_{CO}\, c_{H2O}} = exp_{10}\left(-2.4198 + 0.0003855\, T + \frac{2180}{T}\right) \,. \tag{2.30}$$

When methane is present, it can be included into the reaction system by any reaction that involves it alongside components that are already part of the system.

The dry reforming or steam reforming reaction of methane are possible reactions in this case. Here, the steam reforming reaction was used. It's reaction equation is

$$CH_4 + H_2O \longleftrightarrow CO + 3H_2\,. \tag{2.31}$$

The steam reforming reaction goes to completion at high temperatures. Methane conversion of more than 90% is reached at a temperature above $1000\,K$. The equilibrium constant $k_{steam-ref}$ can be approximated according to [76] using

$$k_{steam-ref} = \frac{c_{H2}{}^3\, c_{CO}}{c_{CH4}\, c_{H2O}} = 4.225 \cdot 10^{15}\, e^{\frac{28879}{T}}\,. \tag{2.32}$$

The equilibrium constant for different temperatures and the resulting equilibrium conversion X of the steam reforming reaction is also shown in fig. 2.12.

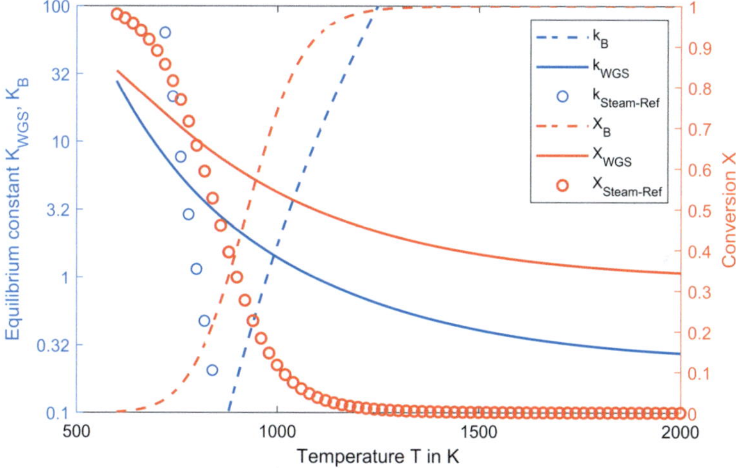

Figure 2.12: Equilibrium constants of the Boudouard reaction, k_B, water-gas shift, k_{WGS} and steam reforming, $k_{steam-ref}$ calculated according to equations (2.28), (2.30) and (2.32). Resulting equilibrium conversions X_B, X_{WGS} and $X_{Steam-ref}$ are also illustrated.

2.3   Electrical

The third range of topics that needs to be covered is electrical engineering. First, some common plasma sources are presented in section 2.3.1. To generate ambient pressure plasma, current limiting topologies are advisable. Two are briefly discussed in section 2.3.2 and 2.3.4. To ignite a discharge, a significantly higher voltage is necessary than for operation. Ignition circuits are presented in section 2.3.5.

The driver circuit supplies power to the discharge. It has an intimate relationship with the discharge: It is well known, that the operation point in a system with a generator and a load can be found as the intersection of both V/I-curves [77]. The same connection applies to the plasma (load) and driver circuit (generator), but as we saw before the V/I-curve of an ambient pressure discharge is not a simple linear one. The driver circuit must be able to reach a stable point of operation, where the desired voltage or current is supplied to the discharge. A good reference example is the V/I diagram of a resistive load coupled to a DC source with a constant output resistance as shown in fig. 2.13.

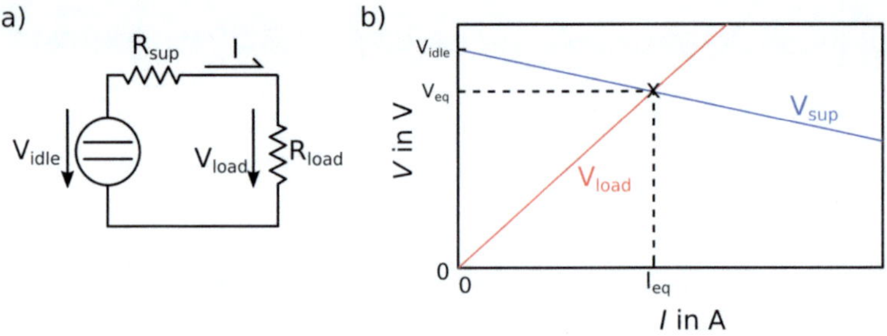

Figure 2.13: Equivalent circuit and V/I diagram of a resistive load and unregulated power supply. The intersection corresponds to output voltage and current of the source.

It is apparent, that the linear slopes give a well defined point of operation at the point of intersection of both curves. This is true for any system of a supply and load, where the $\delta I/\delta V$-slope of the load is greater than that of the supply [78]:

$$\frac{\delta V}{\delta I}_{load} > \frac{\delta V}{\delta I}_{supply}. \tag{2.33}$$

$\delta I/\delta V$ is equivalent to the series resistance with negative sign in the case of the simplified DC power supply and equivalent to the load resistance in the case of a linear load. In a stable operation point, fluctuations in load resistivity or voltage lead to small changes in the current. These changes are smaller, the steeper the angle of intersection of both curves is. The same cannot be said for a ambient pressure discharge driven by a constant voltage power source, which is illustrated in fig. 2.14. A point of intersection of a constant voltage with the V/I curve of a discharge is not always well defined; either there are multiple or none. Operation point A in fig. 2.14 is still stable, but operation point B cannot be achieved by the power supply: The slope criterion given by equation (2.33) is not fulfilled. This manifests in practical experiments: At a given voltage, infinite current could be predicted to flow. Since a plasma cannot be 'destroyed' by high power input, the driver circuit will fail first. A very large current flows in a low resistance arc leading to heating and failure of electrical components. Thus, constant voltage power supplies cannot be used. On the contrary, a well defined point of operation can always be reached with a current limited power supply, which is also shown in fig. 2.14. A constant current can be achieved by using a high voltage power supply in series with a large resistor or dump load, which leads to stable operation in point B. Since dump loads waste a lot of electrical energy, several methods of current control are pursued for the creation of plasma for catalysis, which will be reviewed in section 2.3.2.

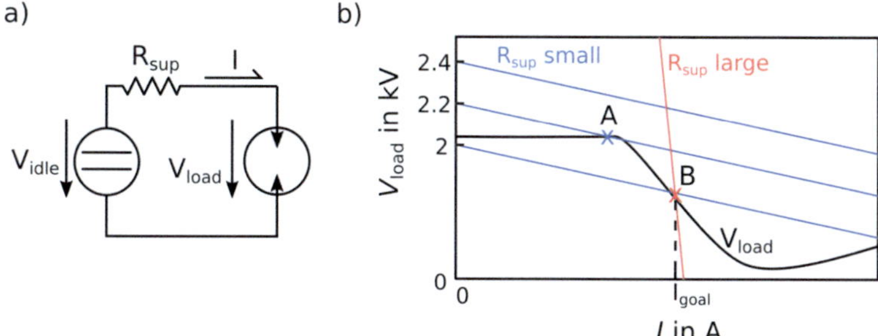

Figure 2.14: Equivalent circuit and V/I diagram of a gas discharge and an unregulated power supply. Multiple operation points are shown at different supply voltages V_{idle} using large and small series resistance R_{sup}.

The stable operation of the plasma is the first challenge, the second one is its ignition. The breakdown field strength of the utilized working gas must be reached in order for a visible (meaning glow or arc) discharge to form. The field required for CO_2 is around $E_{breakdown,CO2} = 3\,kV\,mm^{-1}$ [79]. For a reactor with meaningful size, an ignition voltage of tens of kV is thus necessary. Ignition methods and high voltage generating topologies are reviewed in section 2.3.5.

The third major challenge is presented by the efficient operation of the driver circuit and plasma. Cost of electricity are a major part of operational costs in plasma catalytic applications, since in an endothermal process the reaction enthalpy must be provided by the driver circuit. Efficient electronics also reduce the environmental footprint of a given process, since a large share of carbon emissions can be attributed to electricity. For example, electrical energy consumed in Germany in 2020 can be associated with CO_2 emissions of $\approx 400\,g_{CO2}\,kWh^{-1}$ [80]. A CO_2 negative process such as CO_2 splitting can quickly become a net emitter due to these emissions. The break even energy input for CO_2-splitting can be calculated to be just $396\,kJ\,mol^{-1}$ or approximately $18\,J\,SCC^{-1}$.

2.3.1　Plasma Sources

This section gives a short overview for the most common plasma sources. A plasma source includes a electrical power converter like a driver circuit and the electrode assembly. Plasma sources may be divided by different criteria: Non-thermal or thermal plasma, high- or low pressure operation, self- and non selfsustained plasma, coupling mechanism, excitation frequency and more. Selfsustained means, that the charge carriers that define a plasma are created by the discharge itself, e.g. through emission from the cathode. Non-self sustained discharges rely on an external source of electrons [36].

Selfsustained discharges can be further distinguished by the electrode setup. Bare metal electrodes allow a continuous current to flow, thus they can be used at any frequency including DC. An example of such a cathode sustained discharge is the gliding arc discharge illustrated in fig. 2.15 b) [48]. If a large volume of superheated gas has to be generated, plasma torches are often employed. A high velocity gas flow carries the plasma out of a nozzle, as seen in fig. 2.15 c) [48]. On the contrary, magnetically or dielectrically coupled plasma cannot be excited using DC. Electrons are excited by a temporally changing electric or magnetic field. They are released from the gas molecules inside the discharge, no emission from the electrodes is necessary. Examples of capacitively coupled plasma sources are radio frequency (RF), DBDs and Microwave plasma (MW) [48]. DBDs can operate even as low as 50 Hz up to several hundred kHz. Low pressure RF plasma used in the semiconductor industry requires radio frequency excitation, while microwave plasma sources per definition use microwaves generated by a magnetron. Common frequencies are $2.45\,GHz$ and $915\,MHz$, which have sufficiently large wavelengths to produce large volume plasma and are free to operate in the ISM bands ('Industrial, Scientific and Medical'), which is a part of the frequency spectrum defined by the international telecommunication union [81]. An example of a microwave plasma generator is shown in fig. 2.15 a) [48]. The defining property of capacitively coupled plasma is, that the electric component of the field induces the majority of the energy into the plasma. When the magnetic field is used instead, the plasma source is referred to as inductively coupled. An air coil is used to introduce a strong magnetic field into a stream of gas, typically the plasma is contained in a glass or ceramic tube. An example of this setup is shown in fig. 2.15 d) [48].

2.3.2 Current Limiting

Various configurations exist to create stable discharge at ambient pressure. Regardless if a glow or arc discharge is fed, limiting of the output current is inevitable. Different approaches exist to achieve this, which are illustrated in fig. 2.16. One possibility is to employ a ballast resistor, as shown in a). If the resistor is large enough, a steep V/I-curve can be created [82]. A large inductor can be used instead when alternating current is employed, which is illustrated in b). The current rise is thereby limited to a maximum in each period of operation of the driver circuit. A water electrode handily combines a heat sink with a ballast resistance, since the water acts as a large resistor c) [83]. Instead of an inductor, a capacitor could also be used. A DBD can be viewed as a variation of this approach; the dielectric serves as a capacitor. This limits the amount of charge that can flow through the discharge in each period. DBDs can be built in coaxial d) or planar e) arrangement [48]. Since ballast resistors result in huge energy losses, and inductors and capacitors only can be used with size limitations and in AC plasma sources, another solution could gain popularity over the last years: The use of high voltage capable power electronics. A variety of circuits are able to provide constant current output over magnitudes of

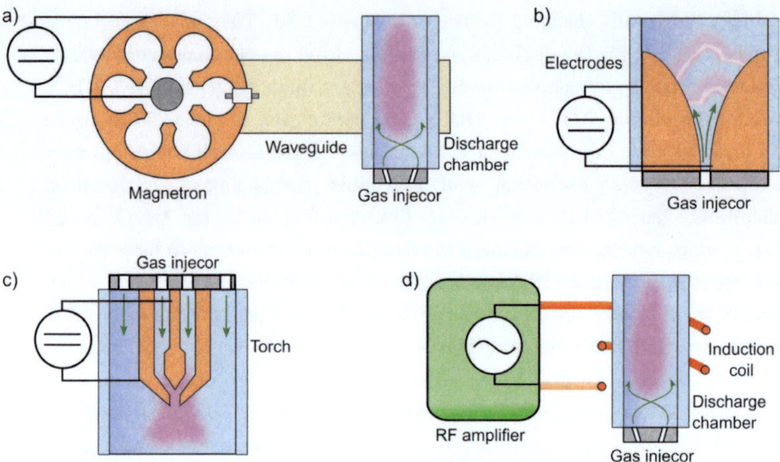

Figure 2.15: Various plasma sources: a) shows a microwave plasma source, consisting of magnetron and waveguide b) illustrates a gliding arc plasmatron with linear electrodes. c) shows a plasma torch. d) pictures an inductively coupled radio frequency plasmatron. All images adapted from [48].

operation voltage f). Omitting the cathode as an electron emitter and instead using a controllable electron generator can also produce stable discharges using constant voltage. The electron emitter is frequently realized as a micro hollow cathode discharge, as shown in g) [84].

As established previously, gas discharges at ambient pressure typically have a negative differential resistance. Since power electronics operate at a fixed frequency during which they either are highly conductive or not at all, a choke inductor is needed to limit the current flowing into the discharge during the 'conductive' phase of each period. The necessary dimension of the inductor can be derived: It must be large enough so that current ripple $\hat{i}$ during each period of operation does not exceed a small fraction of the overall current $I_{discharge}$, which can be calculated as

$$\hat{i} = \frac{T_p V}{L_{choke}} << I_{discharge} \tag{2.34}$$

with period T_p, voltage V and choke inductance L_{choke}. The inductor must also have sufficient size to prevent oscillation in the output current that can occur due to

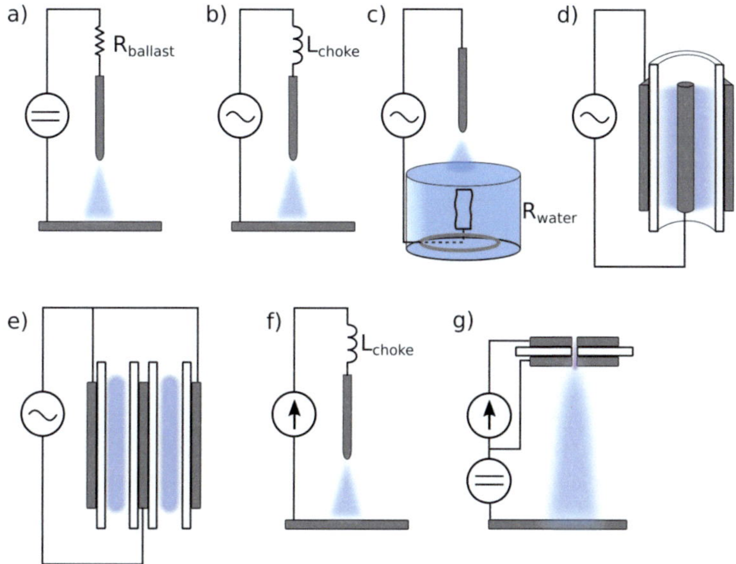

Figure 2.16: Different approaches to current limiting in a gas discharge, image adapted from [13]. a) ballast resistor and voltage source, b) AC source and choke inductance, c) water cathode, d) dielectric barrier in tubular configuration, e) dielectric barrier in planar configuration, f) choke and current source, g) micro-hollow cathode.

hysteresis behaviour of a discharge. It acts as a low pass filter in this circumstance. Any hysteresis device that also features a (stray) capacitance and an inductor can in theory cause oscillation. A cold cathode tube relaxation oscillator is a good example of this behaviour [85]. For the same reason it is important to limit the output capacitance of the driver circuit, since the energy stored in this capacitance can increase instability in the discharge behaviour.

Current limiting topologies can be categorized into two major groups: Passively stable and actively regulated circuits. In an actively regulated circuit, the discharge current is measured and a semiconductor switch is used to control it. This can take different forms, with either one or more switches in a buck-, boost- or forward converter, that can also include a transformer. For the generation of very high voltages, a transformer is usually necessary since semiconductor switches with a blocking voltage of more than a few kV are prohibitively expensive. An example topology for the generation of high voltage direct current using active regulation is shown in fig. 2.17.

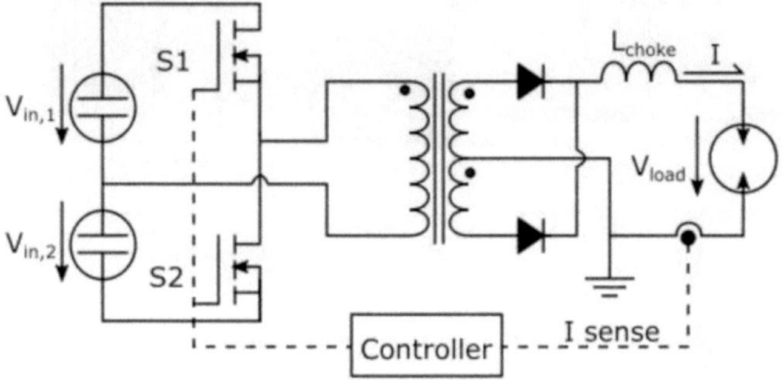

Figure 2.17: Example of an actively regulated power supply used as a plasma driver.

Passive stability omits the need for high frequency measurement and control of semiconductor switches. The switches can instead be operated at controlled frequency, or using feedback measurements as in a bi-stable oscillator. Passive stability in this case means, that a constant output current is maintained over a wider range of output voltages. An LLC-converter is an example for a current limiting topology. Since the plasma driver proposed in this thesis uses a similar topology, it is covered in more detail in the following section.

2.3.3 Royer and Baxandall Oscillator

'Royer osciallator' is a term today used to describe a variety of resonant converter circuits. In the original invention, a LC-parallel circuit is fed at resonant frequency by transistors, which are driven by a feedback winding on the same transformer [86]. The design originally produced square wave output voltage and was adapted by Baxandall to generate sine waves by changing the wiring of the transformer [87]. A more efficient version uses two MOSFETs, which in turn are feedback controlled by a second pair of MOSFETs that generate the switching signal. A choke inductor provides constant current to the oscillator circuit. The circuit is zero voltage- and current switching and can operate with very small losses. The resonant circuit reaches an inherent gain of $G_r = \pi$. A transformer is used to increase the voltage further. The circuit is not current limiting by itself, but it can be made to be by two measures: An additional inductor can be placed in series to the transformer to serve as ballast, or a transformer with high stray inductance can be used. Furthermore, a flyback configuration can be realized by using a single diode rectifier on the high voltage side. Combining both measures yields a plasma driver with nearly constant transconductance. A circuit diagram of a royer oscillator is shown in chapter 3 in fig. 3.5.

2.3.4 LLC-Converter

LLC-converters can achieve high efficiency and a compact size with moderate complexity [88]. They have been employed before because they can achieve good current limiting behaviour [89]. The LLC- converter uses a series resonant circuit in series to a transformer, that is fed by semiconductor switches in half bridge or full bridge configuration. 'LL' refers to two inductances, while 'C' refers to a capacitance in this circuit. One of the inductances is the magnetizing inductance of the transformer L_m. It describes all non-coupled inductances of the transformer, e.g. inductances in which a primary current does not necessitate a secondary current. The equivalent circuit for a transformer is given below in fig. 2.18 [88]. To understand the LLC circuit, the schematic of a non-ideal transformer must be used. R_p and L_p are the resistance and leakage inductance of the primary coil. R_s and L_s represent the same values for the secondary. R_{Fe} represents the core losses and L_M the magnetizing inductance of the core. The winding ratio of the transformer a is

$$a = \frac{N_2}{N_1}.$$

(2.35)

The square of the winding ratio a is used to translate the inductance L_s and resistance R_s of the secondary coil to the primary side:

$$L'_s = L_s\, a^{-2}$$
$$R'_s = R'_s\, a^{-2}.$$

(2.36)

The resonance inductance L_r and resonance capacitor C_r are placed in series to the transformer. When the leakage inductance of the transformer cannot be neglected, it it adds to L_r. They form the resonant circuit, also called tank circuit. This circuit is then fed by switches in a half- or full bridge configuration. On the high voltage side, a rectifier creates a DC-current from the amplified AC-voltage.

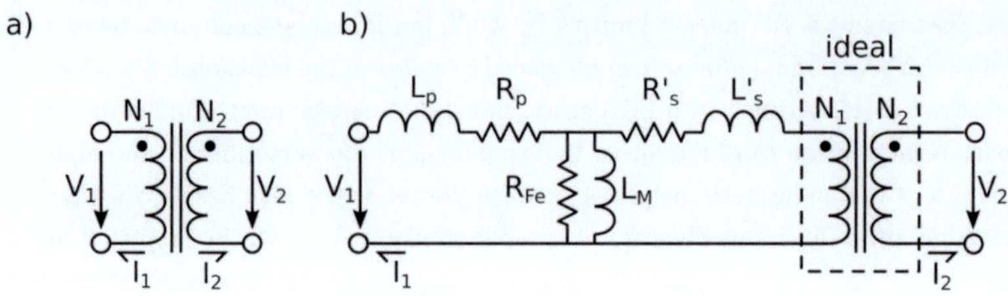

Figure 2.18: Equivalent circuit models of a real (a) and ideal transformer (b).

Fig. 2.19 gives an overview of the whole assembly, highlighting two important cases: When no load is present, the transformer can be treated as an infite resistance, so all current on the primary side has to flow through the capacitor and both inductances. In a secondary shortcut, representing a high load, the transformer instead is treated as a short circuit. All current only flows through L_r and C_r.

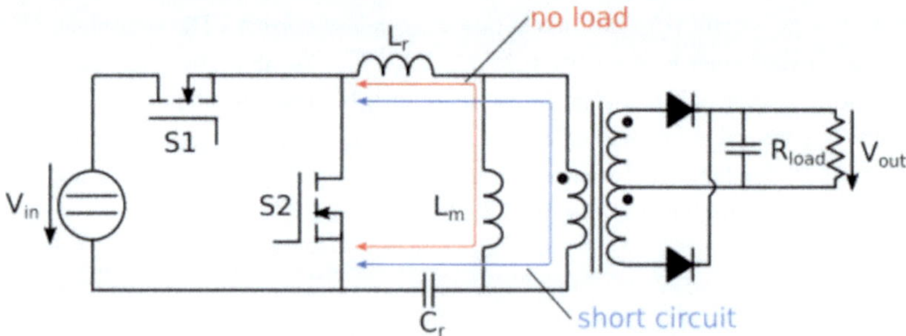

Figure 2.19: Circuit diagram of an LLC converter in half bridge configuration. Current paths at no load condition and infinite load are indicated [88].

These two states lead to two distinct resonant frequencies, $f_{r,load}$ and $f_{r,idle}$. They can be calculated according to [88, 90, 91] using

$$f_{r,load} = \frac{1}{2\pi L_r C_r}$$
$$f_{r,idle} = \frac{1}{2\pi (L_r + L_m) C_r}. \tag{2.37}$$

The output characteristics of an LLC converter are shown below in fig. 2.20 [88, 90, 91]. Mostly, LLCs are actively controlled by frequency modulation at constant duty cycle. The frequency can then be used to regulate output voltage or current. However, the LLC can also be driven at self-resonance frequency using feedback. At the lower $f_{r,noload}$, all energy introduced into the circuit cannot pass through the transformer, thus it builds up in the resonant circuit. The total voltage gain G_r can be in the range of

$$G_r(Q) = \frac{\hat{V}_r}{V_{sup}} = 0...5 \tag{2.38}$$

in practice, depending on quality factor Q of the resonant circuit. The quality factor describes the ratio of characteristic impedance and load resistance. It can thus be calculated using the load resistance R_{load}, L_r and C_r [90, 91]:

$$Q = \frac{\sqrt{L_r/C_r}}{R_{load}} \tag{2.39}$$

The gain factor G_r does not include the step up transformer ratio a, so the actual maximum output voltage V_{out} is

$$V_{out,idle} = V_{sup}G_r\,a. \qquad using\,a = \frac{N_2}{N_1} \tag{2.40}$$

in a full bridge, or push-pull configuration [90, 91]. In a half bridge, the output voltage is also halved. When a high load is present, the resonance frequency shifts up to $f_{r,load}$. The resonance circuit is drained by the transformer, so the gain of the resonant circuit drops to $G_r = 1$. In this operation point, the LLC converter displays constant output voltage at arbitrary current. The transconductance g_m of the circuit, described by

$$g_r(Q) = \frac{I_{out}}{V_{in}}. \tag{2.41}$$

becomes very large, depending on the quality factor of the circuit. In between and outside of both operation points, the values of G_r and g_r change in a non-linear way, which is illustrated in fig. 2.20.

To achieve constant voltage, the LLC converter can be operated near the upper resonance frequency $f_{r,load}$ or above of it. The impedance of the series resonant circuit Z_{in} is zero at the higher resonance frequency f_{load}, thus the whole input voltage is applied to the load.

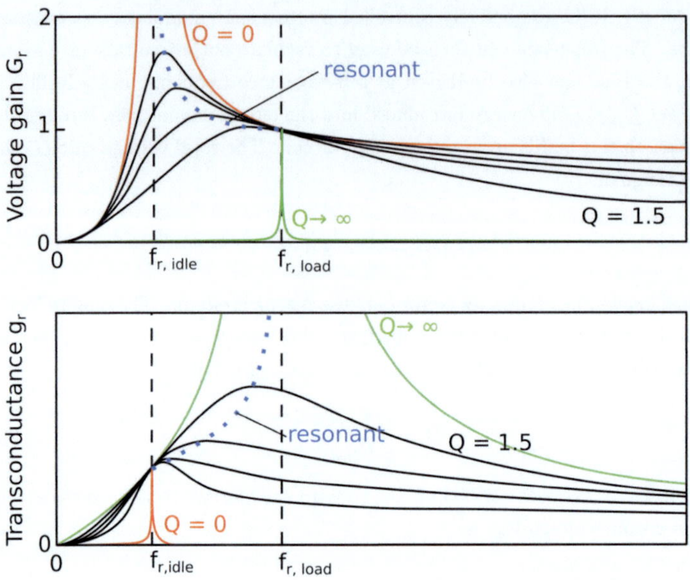

Figure 2.20: Transconductance g_r and voltage gain G_r of the LLC-converter at different frequencies and quality factors. Operation points using resonant switching are indicated [90].

On the other hand, current limiting is achieved at $f_{r,idle}$. However, switching might transition from zero voltage (ZVS) to zero current (ZCS), which is accompanied by higher losses. The impedance of the series resonance circuit is

$$Z_{res} = \sqrt{(X_L - X_C)^2} = \sqrt{\left(2\pi f_{r,idle}L - \frac{1}{2\pi f_{r,idle}C}\right)^2}. \qquad (2.42)$$

The maximum output current $I_{shortcircuit}$ at $f_{r,idle}$ is limited by the impedance

$$I_{short}(f_{idle}) = \frac{V_{sup}}{Z_{res}}. \qquad (2.43)$$

This equation predicts that current limiting can be achieved at every frequency except resonance f_{load}. The short circuit current can be calculated depending on the frequency. Since the voltage gain at resonance frequency can be very large, which leads to high magnetization currents, the circuit should best be operated slightly above its lower resonance frequency. Otherwise, the transformer might saturate, leading to additional losses. A trade off must be made between current limiting capability and magnetic losses at each frequency.

The combination of high voltage gain at no load and current limiting behaviour at high load make the LLC converter ideally suited for the use as a plasma driver. The output voltage drops off quickly after ignition. However, the primary voltage is limited by the capability of the semiconductor switches. Also, very excessively large transformer ratios a are not practicable. An additional voltage amplifier is thus needed for ignition of the plasma. Ignition mechanisms are covered in the following section.

2.3.5 Ignition Methods and Circuits

The easiest approach to igniting a plasma is to use the two electrodes that are already present in an electrically driven plasma, and apply a high voltage to them. Different circuits allow the generation of ultra high voltage or voltage bursts from a moderate supply voltage. Two options are mainly used in practice: Capacitive voltage multiplier circuit and the ignition transformer.

A capacitive voltage multiplier is an array of capacitors and diodes, that can be scaled to arbitrary length. Different configurations exist: The Greinacher-Cockroft-Walton multiplier is the oldest technology and commonly referred to simply as voltage multiplier circuit. Modern variations include the Dickson-Charge pump, Marx-generator (also known as serial-parallel multiplier), Fibonacci multiplier and 2^N-multiplier [92]. Of those five, the Dickson-Charge pump, Fibonacci multiplier and 2^N-multiplier require at least one circuit element to be able to withstand the full output voltage, which is impracticable at very high voltages. The Marx-generator either only works with a discharge gap and only at a specific input voltage, or it requires active switches - both of which is unattractive in this application. This leaves the voltage multiplier or Greinacher-Cockroft-Walton multiplier [92]. Each stage consists of two diodes and two capacitors. The biggest advantage to alternatives such as charge pumps is, that each element only needs to withstand twice the supplied peak voltage $\hat{V}_{sup}$. Each stage increases the output voltage by the amplitude of the applied voltage. The circuit diagram is given below in fig. 2.21 a). The output voltage is fully rectified. The current that can be sustained for a given ripple voltage is proportional to the capacitance C_{mult} and frequency of the applied voltage. One drawback of this topology is, that the diodes in lower stages are subject to a high current. The current depends on the number of stages N_{stages} and output current I_{out}. The current through the first diode is approximately

$$I_{in} = I_{out} N_{stages},\tag{2.44}$$

while the output voltage without load can be calculated as

$$\hat{V}_{out} = 2\hat{V}_{sup} N_{stages} - \delta V\tag{2.45}$$

with voltage drop δV due to load current [92]. The high supply current I_{in} is a limiting factor for application, especially since very high voltage ultra fast diodes are required, which are usually not able to handle large currents. The voltage multiplier has a linear V/I-curve, where the drop in voltage δV can be approximated according to [92] using

$$\delta V = \frac{I_{out}}{f\,C_{stage}}\frac{2}{3}N_{stages}^3\tag{2.46}$$

with frequency f and capacitance of each capacitor C_{stage}. An additional voltage ripple occurs, which is not considered in this application.

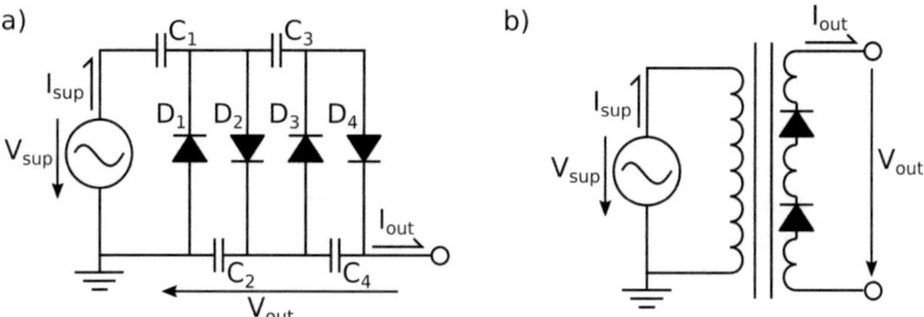

Figure 2.21: Circuit diagram of a voltage multiplier with two stages a) and a diode split transformer b).

An alternative is the use of ignition coil or transformers. An ignition transformer typically has a very high transformer ratio, in excess of $a > 30$, and also high impedance. This makes it short-circuit proof, which is important since an arc created on the output may have very low resistance. An ignition transformer is driven by a continuous alternating current or operated using short pulses of ultra high current, that need to be switched off quickly. This means, it can operates as flow converter or as flyback converter.

To achieve large transformer ratios and conversely high voltages, a high secondary inductance is necessary. This limits the current throughput. The problem can be solved by using a diode split transformer instead. A diode split transformer consists of several winding packages connected in series, where a diode is placed in series to each coil package. The output voltage is thus direct current, which means a capacitor can be charged to allow for high pulses of output current. An advantage of the diode split transformer is that each diode only needs to conduct a current of

$$I_{Diode} = I_{out} = \frac{I_{sup}}{a}.$$
(2.47)

The output voltage is

$$V_{out} = a\,V_{sup}.$$
(2.48)

Generally, a diode split transformer is less demanding on the diodes employed, but requires a well insulated transformer. It needs to withstand the ignition voltage, which is up to $50\,kV$ insulation voltage in the power supply constructed in the scope of this thesis. It must be noted, that the driver circuit and ignition circuit operate the same two electrodes in both setups. This can be a drawback, since additional diodes are necessary to prohibit the current from the ignition circuit to flow back into the driver circuit. This is also illustrated in fig. 2.22 a).

For this reason, arc welders use a different approach to ignition: An air core transformer is placed with its secondary in series between driver circuit and electrode. A high frequency alternating voltage is then induced into the coil. This voltage is sufficient to ignite a plasma, but can also easily be blocked with a filter capacitor instead of a diode. Since a choke inductance is required in many efficient plasma drivers anyway, this choke can be built with a second winding with low turn number for ignition. This setup is illustrated in fig. 2.22 c).

The limitations of a two electrode setup can be overcome in part by using a third electrode. It can be biased towards one of the others with a high voltage. Depending on the spacing, different methods of ignition are possible. The third electrode can be covered in a dielectric and placed near the cathode. It can be operated as a DBD, which creates a sufficient charge carrier density to reduce the ignition voltage necessary for the main discharge. This setup is illustrated in fig. 2.22 b). It was briefly considered for use in the reactor built for this thesis, but discarded due to

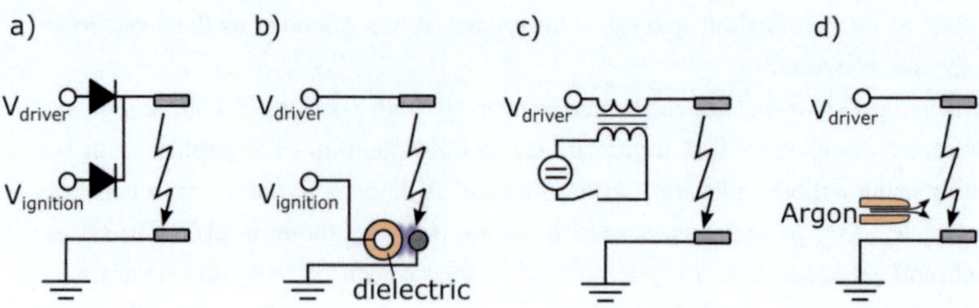

Figure 2.22: Circuit diagram of a two electrode ignition system a), three electrode ignition system using DBD b), high frequency or pulse transformer c) and inert gas aided ignition using argon d).

the additional complexity. The third electrode can also be carried out as a micro hollow cathode. The advantage of this setup is, that the micro hollow discharge creates charge carriers while displaying a current limiting behaviour by itself [84]. A third possibility is the heating of the cathode to easier emit charge carriers.

Instead of trying to ignite the working gas directly, a gas with a lower breakdown field strength can be used, which is illustrated in fig. 2.22 d). Plasma sources have been previously demonstrated that inject argon or helium to ignite the plasma, which results in a up to 20 times lower ignition voltage, e.g. [93]. In addition to ignition systems, an electrode setup with variable electrode distance can be used to reduce the necessary ignition voltage [58].

Chapter 3

Experimental Methods

The experimental findings presented in this work are generated with successive experiments. Main focus are the generation, characterization and application of the DC-discharge. The results were previously published [94, 95]. In addition to the experiments, the boundary conditions, assumptions and methods for simulations that were ran as a party of this dissertation are given in this chapter.
All experiments used similar equipment which is discussed in detail in this chapter. While the reactor and driver circuit were scaled, the periphery remained similar between the studies. The consecutive versions of the plasma reactor experiment will be referred to as mk 1 and mk 2.

A diagram of the mk 1 reactor design and experimental setup is given in fig. 3.1. In a), a cross section of the reactor is shown. The ring electrode consists of a stainless steel pipe section. The insulator is made from aluminum oxide. Neodymium ring magnets are used with a remanence field strength of $B = 1.2\,T$. Teflon tubing is used for the gas. Every transition is sealed using solvent free silicon compound, which was applied form the outside. b) shows the dimensions of the reactor. The diameter of the pin electrode is 3mm. This means, that the discharge gap is between 6.5 and $8\,mm$, depending on the position of the discharges origin on the electrode. The pin electrode can be adjusted to raise its tip out of the discharge plane. In c), the assembled experiment is illustrated. The reactor is held by a PVC mount, which is attached to a stand (not shown). Details about the assembly can be found in the following subsections.

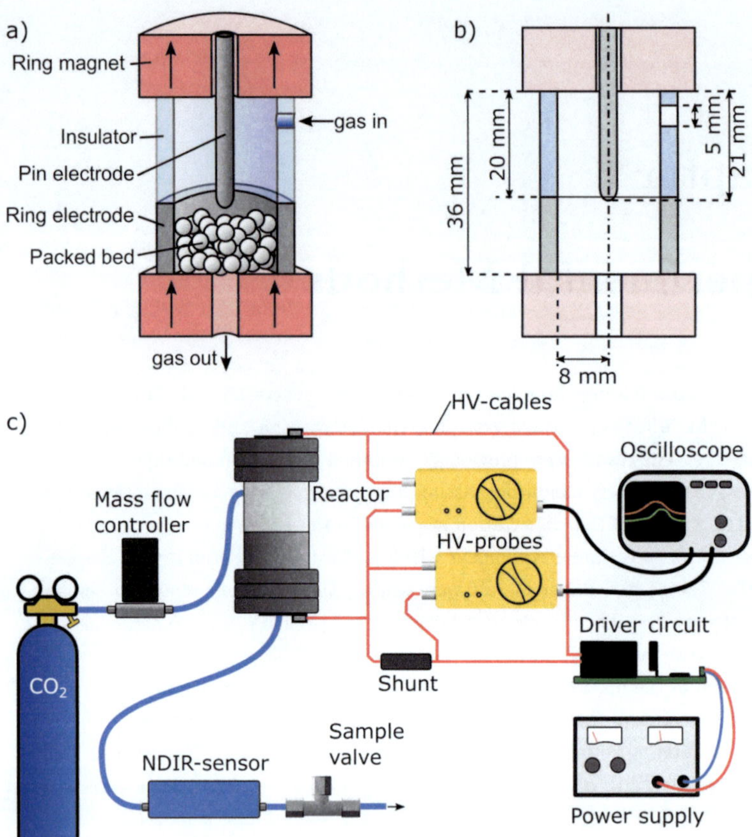

Figure 3.1: Experimental setup of the mk 1 plasma reactor. a) shows the cross section of the reactor. In b), the dimensions of the reactor are given. c) shows a schematic of the assembled experiment for measuring the CO_2 splitting performance.

In the mk 2 assembly, some changes were made. The automated mass flow controller is replaced by an analogue rotameter with a maximum flow of 1.25 SCCM of CO_2 since it is limited to 200 SCCM. The gas flow is adjusted using a needle valve. The assembly was again calibrated using a displacement cylinder. The gas analysis is the same, consisting of NDIR-sensors (Non Dispersive Infra Red) for CO and CO_2. The assembly is shown in fig. 3.2.

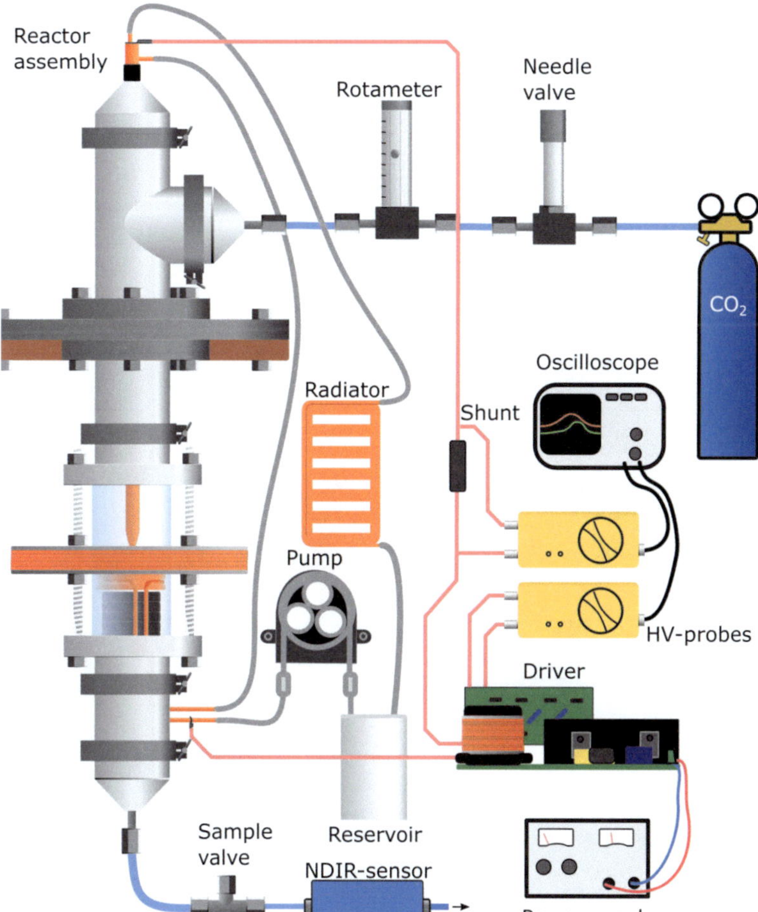

Figure 3.2: Experimental setup of the mk 2 plasma reactor. Not shown are the auxiliary power supply for the driver circuit and the auxiliary power supply used to feed the Helmholtz coils.

The electrodes are actively cooled. A peristaltic pump is used to circulate silicon oil through the ring electrode, pin electrode, a copper radiator and a reservoir. The driver is fed by a lab power supply. Two differential high voltage probes and a shunt are used to measure applied voltage and current. All components are mounted into a Acrylic glass enclosure, which is connected to an exhaust ventilation system. A cross section of the reactor is shown in fig. 3.3 a).

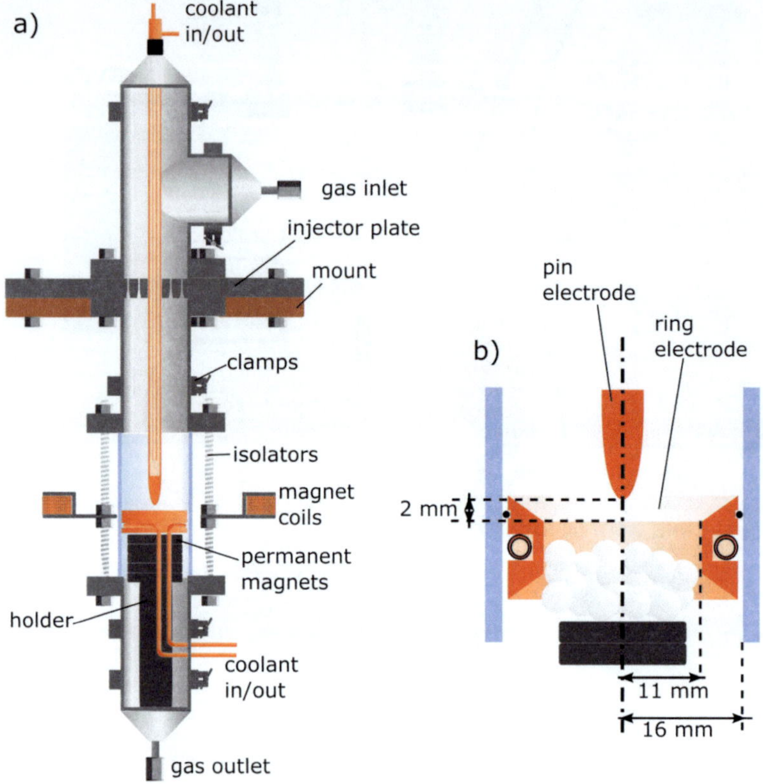

Figure 3.3: Detailed view of the mk 2 gliding arc reactor. a) shows a cross section of the reactor assembly. In b), the dimensions of the electrodes are given.

The pin electrode features a coaxial structure which allows cooling oil to circulate through it. The ring electrode has a groove on the outside where a copper pipe is mounted for cooling as well. The gas inlet is positioned in a T-pipe on top.

The working gas flows through a custom injector plate, which has 22 axial nozzles arranged in concentric circles to achieve uniform gas flow. The injector plate is positioned 120 mm above the discharge plane. All metal parts except the electrodes are made from 304 nonmagnetic stainless steel. The pipe section surrounding the electrodes is made from quartz glass. The coolant pipe of the ring electrode exits the reactor below the discharge plane. Neodymium magnets can be placed below the ring electrode to generate magnetic fields of up to $50\,mT$, which are naturally non-uniform. The magnets are held by a holder. Zirconia balls with a diameter of $3\,mm$ are placed on top of the magnets. The rign electrode is thus filled up to the discharge plane. Magnet coils can be used to generate the axial magnetic field. Note, that only one magnet coil is illustrated. In fact, two coils were used in a Helmholtz arrangement. The effective diameter of the coils is $70\,mm$, their distance is $50\,mm$ with the discharge located in the centre. The coils have a resistance of $1\,\Omega$ and are able to generate a field of $32\,mT$ at a current of $10\,A$.

Clamp joints are sealed using high temperature silicon compound and silicon O-rings. Screw joints are sealed using copper gaskets. Gas inlet and outlet are VCR-pipe mounts (Vacuum Coupling Radiation, a type of coupling that can withstand vacuum and positive pressure), which are sealed with silver gaskets. The dimensions of the electrode assembly are shown in fig. 3.3. The pin electrode has a diameter of $6\,mm$. Its tip is located $2\,mm$ above the discharge plane. The inner diameter of the ring electrode is $22\,mm$. The inside diameter of the glass pipe section is $32\,mm$. The glass section is compressed between two O-rings and additionally sealed with silicon compound. Polymer threaded bolts are used to compress the glass pipe, while at the same time serving as insulators. The upper and lower portion of the reactor are not electrically insulated but instead correspond to cathode respectively anode potential. This makes it especially important to import the reactor in a safe enclosure.

3.1   Electrical Assembly

Each plasma reactor is fed by a driver circuit. Different kinds of driver circuits were used throughout the experiments. Mk 1 is fed by a flyback converter. Mk 2 uses a resonant converter. The mk 3 driver circuit, which uses an LLC-converter could not be finished in time. These drivers were picked because they each represent the most sensible solution in their respective power range. In turn, the drivers are powered by a lab power supply. Initially a $35\,V$ supply was used, which was substituted for a more capable one since higher voltages were needed for the mk 2 driver. The supplied direct current is fed into the driver circuit.

3.1.1 Mk 1 Driver Circuit

The mk 1 driver circuit is based on a self-switching oscillator. It generates sine wave voltages at $f = 50\ldots100\,kHz$. Since it is operated in its natural frequency, the oscillator provides an output voltage gain of $g = \pi$. This is further amplified by a step up transformer. The transformer uses a high stray inductance and is operated as flyback transformer. It can thus reach very high output voltage. A voltage multiplier is fed by the secondary of the transformer, which produces direct current at its output. Between 1 and 5 stages are used, depending on the required ignition voltage. The assembly is shown in fig. 3.4.

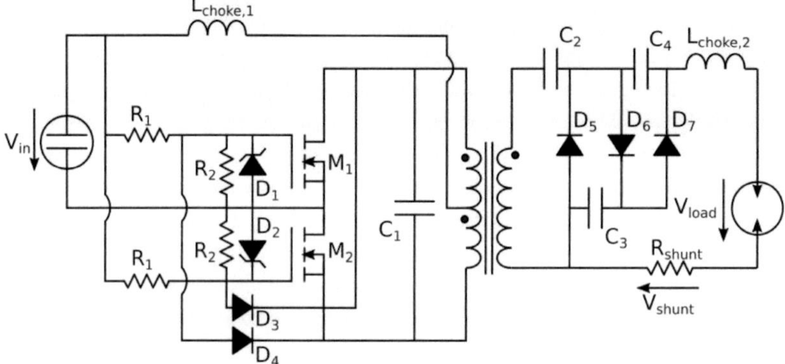

Figure 3.4: Circuit diagram of the driver circuit used to drive the mk 1 plasma reactor.

3.1.2 Mk 2 Driver Circuit

The mk 2 driver circuit was implemented to overcome limitations of the first version. A resonant converter features higher power capability with a transformer core of the same size. Additionally, the setup of the oscillator as a Royer/Baxandall Converter allows for faster switching times and thus less switching losses on the MOSFETs that are used. On the high voltage side of the transformer, another obstacle needs to be overcome. As discussed in chapter 2.3.5, the current though the first diode in a multiplier circuit is equal to the stage number n multiplied by the output current I. This would lead to large dimensions for the diodes that are used. Diodes that can handle the high voltages (20kV), currents (> 1A) and frequencies (100 kHz) at the same time are rarely available and expensive. For his reason, a single diode rectifier is used in parallel to the multiplier. The multiplier is connected in series with an additional ballast resistor of $20\,M\Omega$. The operation current after ignition thus flows through the single diode resistor instead of the voltage multiplier. The assembly is shown in fig. 3.5.

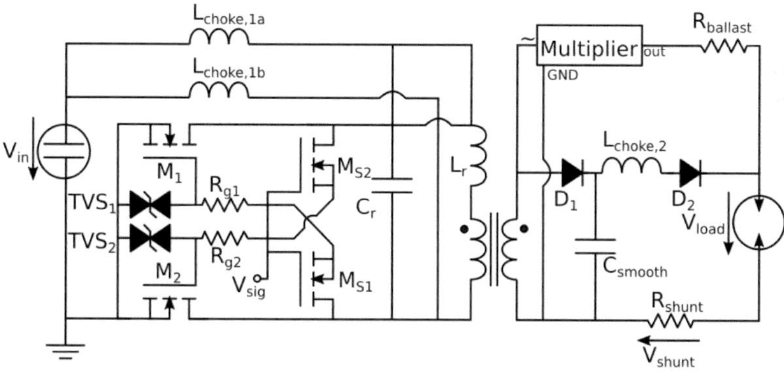

Figure 3.5: Circuit diagram of the driver circuit used to drive the mk 2 plasma reactor.

3.1.3 Measurements of Electronics

The burn voltage in the plasma is measured using differential high voltage probes. Differential probes are required, because the plasma reactor is operated at a floating potential to achieve a more stable discharge with less capacitive coupling. The probes used are manufacture by Cal Test Electronics (CT4079- Differential Probe, 200:1 /2000:1). The discharge current was measured with a similar probe and a low inductance shunt resistor. In mk 1 a $12\,k\Omega$ resistor was used, in mk 2 a $100\,\Omega$ shunt was used. All data was acquired by a tektronix mso2004b oscilloscope.

3.2   Gas Sensing

To judge the performance of the reactors, volume flow and gas composition need to be measured. The input gas concentration is considered constant, since the gas is drawn from cylinders with at least 99.6% purity. Volume flow into the reactor is controlled by a mass flow controller (MKS GM50A013202RMM020). The outgoing flow is not measured, since virtually all flow measurement devices only work when they are calibrated for the working gas. The gas on the reactor outlet has varying composition, which would complicate accurately flow rate measurements.
The mass flow controllers and rotameter are calibrated by a water displacement cylinder. The time to fill a cylinder of known volume can thus be recorded. This measurement is taken at different set flow rates to guarantee proper calibration in each operation point. Temperature of the gas and ambient pressure relative to standard conditions are taken into account. The seal tightness of the assembled experiment is also confirmed by using a displacement cylinder.

The gas composition on the reactor outlet is measured with two independent systems. Non-dispersive infrared sensors use the absorption of specific wavelengths of light by each gas molecule to quantify its abundance, relative to ambient pressure. The sensors can be used in-line and give concentration values with high revolution in time (about 1 value per second). Two *smartgas* 'flow Evo' sensors are used, one for CO and one for CO_2. Both gases have similar absorption lines. Thus, when reading sensor values, crosstalk needs to be taken into account. A linear relationship was used to subtract crosstalk. The value measured by the CO sensor c_{CO} is adjusted by a fit-factor $\beta = 0.01$,

$$c_{CO} = c_{CO,measured} - \beta \, c_{CO2,measured} \, . \tag{3.1}$$

The fit factor is determined by a calibration with pure CO_2, in which the CO sensor is adjusted to give a concentration of $c_{CO} = 0$. Using this formula gave accurate values for concentration in calibration tests. The measurements were confirmed by a gas chromatographer, 'Trace 1310' by *Thermo Scientific*. It uses two ShinCarbon packed columns and an oven that is heated to $180°\,C$ during each measurement. The gas chromatographer is sensitive to hydrogen, nitrogen, oxygen, CO_2, carbon monoxide and methane. It can thus be used to detect air contamination and check if oxygen and carbon monoxide are present stoichiometrically, thus confirming that the reactor is airtight and no side reactions occurred. Samples of $0.5\,ml$ of gas are extracted downstream from the reactor from a sampling valve using a gas tight syringe. These are then fed into the gas chromatographer.

To ensure, that the reactor reached steady state, multiple conditions were met for each measurement:

- The reactor was run for at least twice the gases residence time

- The gas concentrations had reached steady state

- The reactor was run for at least one minute to reach thermal equilibrium

3.3 Plasma Characterisation

To understand the results obtained by experiments, the plasma generated in the reactor has to be characterized. This is done on a very basic level with little requirement for special instruments. Questions of concern are the rotation rate of the discharge channel of the gliding arc reactor, distribution of electric field inside the plasma and power density of the plasma. Electrical and optical properties were studied to learn about them.

3.3.1 Rate of Rotation

As discussed in chapter 2.1, the discharge in the studied reactors forms a rotating filament. The rotation rate was measured with two independent methods: First, a high speed camera which is capable of acquiring up to 960 images per second. This is in theory sufficient to measure period of down to $T_{rot} = 3\,ms$. The burn voltage of the discharge can be used as a second indication: It is dependent on the discharge current and electrode distance. By placing the pin electrode off-centre in the ring electrode, the distance varies within each revolution. This creates a variation in the voltage with the same frequency as the rotation of the filament. Since the burn voltage is measured with very high resolution, rotation periods of less than $T_{rot} = 1\,ms$ could be measured. A drawback is, that the pin placement will have an unknown effect on rotation rate and discharge properties.

3.3.2 Distribution of the Electric Field

To measure the electric field in the discharge, the local voltage at different locations in the plasma is measured. Usually, this would be done by a Langmuir probe. A

Langmuir probe is subjected to a saw-tooth voltage by an external power supply, and the current is then used to calculate electron density, -temperature and local potential. Since no probe was at hand, a very high resistance high voltage probe was constructed instead. A $10\,G\Omega$ resistor is provided with a wire tip on one side, while the other is fixed to a coaxial cable. It forms a 1:10,000 resistive voltage divider with the input resistor of the employed oscilloscope, which is $1\,M\Omega$. This setup allows the wire probe to be placed inside the plasma. A wire probe like this offers the advantage of simplicity. The drawback is, that it distorts the electric field inside the plasma where it is placed. However, the current flowing through the wire probe is low when compared to the current density of the discharge and wire surface area. By moving the wire through the plasma, the potential at different locations can be measured. This can be seen in fig. 3.6. The electric field can then be calculated as the differential of the electric field.

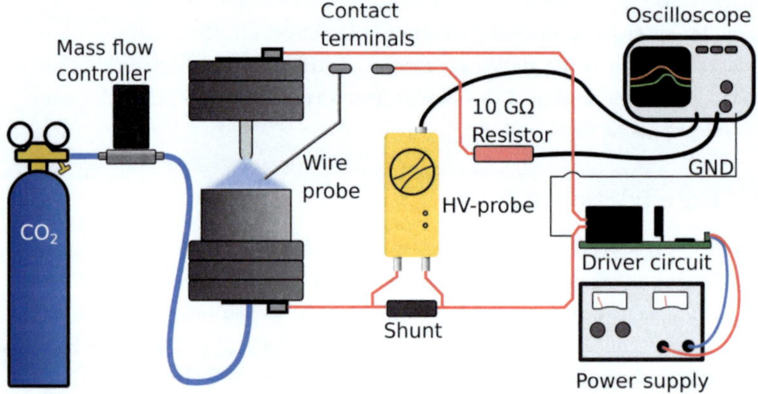

Figure 3.6: Wire probe and its placement for measuring local potential in the mk 1 reactor.

3.3.3   Power Density

The power density of the plasma is calculated from current density and electric field. The electric field can be measured as described in section 3.3.2. To estimate the current density optical observations are made. All the discharge current in a filamentary discharge will flow through the centre of the luminous area due to multiple effects. The luminosity of a patch of plasma can be estimated by taking photographs and high speed video of the plasma with varying exposure time. This method gives a very crude estimation, because the grid on which observations can be made is very coarse. When a channel is visible with a circumference of 1mm, the only certain information is that the majority of charge carriers will flow within this 1mm diameter column, with no extra information on distribution. This limitation will apply to all techniques, hence this method is considered sufficient for approximate results.

3.4 Feasibility Studies

The chapters 5 and 6 investigate methods for carbon monoixde- or syngas production by plasma. Chapter 7 is concerned with economical and ecological feasibility of these plasma processes. They are compared to other approaches, such as electrolysis at low or high temperature. Since most of the worlds productivity is still fossil fuel powered, a comparison must also be made to classical fossil feedstock processes such as steam reforming of natural gas. Focus are energy consumption as well as carbon emissions. An in-depth feasibility calculation would rightfully deserve to be performed in a separate thesis, hence only a simplified calculation is made. Since efficiency and conversion of plasma processes as well as scaling effects are not sufficiently determined yet, an approximate calculation seems appropriate.

A majority of the worlds production of syngas, and hence carbon monoxide, is used for methanol synthesis. Since this is an established process that uses syngas with 2 parts hydrogen and 1 part carbon monoxide, the production of 2:1 syngas will be used for comparison of the technologies. The considered plant capacity is 100 tons of syngas per day in respect to the mass of only H_2 and CO.

3.4.1 Energy and Mass Balance

Syngas plants typically operate within a surrounding network of processes. It is thus hard to view a process that produces syngas and the accompanying emissions in isolation. Heat and steam are exchanged between reactors alongside products. The chain of reactors in the plant is divided in three categories for the sake of simplicity: First, syngas is produced. This step is the main interest in this work. Second, it is refined and purified. This includes CO_2-absorber, condensers and water-gas shift reactors. Third, the remaining syngas is utilized. This chain is illustrated in fig.

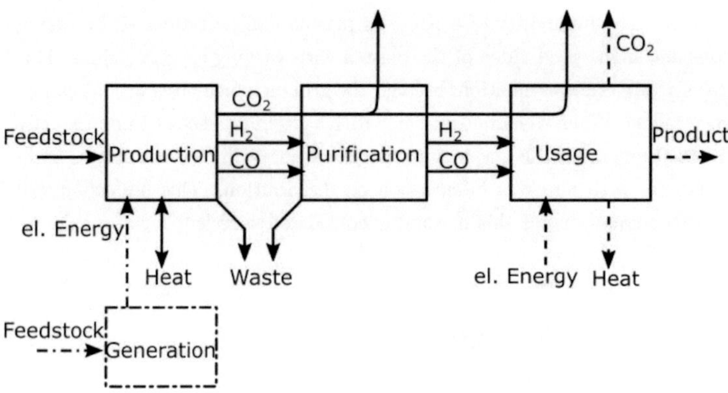

Figure 3.7: Simplified overview of a power-to-X production chain using syngas. Dashed lines are not considered in the calculation, dotted lines describe indirect emissions. Only production processes are included in plant cost.

3.7. Although emissions of subsequent processes are not included, emissions that originate in the original syngas production step and are carried with the product stream need to be included in the CO_2 balance. For example, off-gas or tail-gas, which is the gas that remains after methanol is condensed from a Fischer-Tropsch reactor, is often burned. Otherwise, only the production and purification process is considered in plant costs and material costs. Removal of compounds that are not a general problem to methanol synthesis in theory such as CO_2 or steam is not considered.

The costs of producing any product are comprised of two main factors: Capital expenditures (CapEx), and operational expenditures (OpEx). Capital costs include the investment into a plant. They are repaid as an annuity over a typical span of 10 years and in this time add to the product cost. The costs per year are calculated using the annuity factor method based on a fixed interest of 5% per year, which gives an yearly annuity factor of $a = 0.1295$.

Plant costs are estimated based components of similar plants found in literature. All components shown in process diagrams are included in the CapEx and also listed in the according table. Costs Γ for typical components were adapted from literature where similar plants are investigated and adapted by scaling to the required nominal power or flow. Cost scaling is assumed exponential with a 15% decrease in specific costs per decade. This empirical value is meant to represent a medium economy of scale effect; considering individual scaling factors for different components would go beyond the scope of this thesis. E.g. costs for each component Γ_i scaled by feed mass flow $\dot{M}$ are calculated by

$$\frac{\Gamma_i}{\Gamma_{ref}} = \left(\frac{\dot{M}_i}{\dot{M}_{ref}}\right)^{0.85}. \tag{3.2}$$

When no similar components were available, the costs were estimated using McGraw Hill's online tool [96], which in turn is the same data base used in DWSIM, a well-established chemical process simulation tool. For good measure, an extra 10% of all component costs were assumed as overhead for piping and installation, another 20% for controls. Thus, yearly CapEx are

$$CapEx = a\,1.3 \sum_i \Gamma_i. \tag{3.3}$$

An exception to this are the costs for the plasmatron. The costs for a $3\,kW$-Prototype must be determined in detail, since no data is available for scaled non-thermal gliding arc reactors in industrial use. A 3 kW reactor is under construction as part of this work at the time of this writing, manufacturer pricing for bulk orders of 100 units were used. Added to this are costs for assembly, unit testing and a 10% overhead. Costs for plasmatrons are also scaled according to equation (3.2). A cost lineup can be found in appendix C. OpEx for a chemical plant include maintenance and repairs, personal wages, electricity and other energy sources, feedstock chemicals and, since the focus lies on the EU, CO_2 emission certificates. Maintenance is considered as a fixed percentage of CapEx of 4% per year, unless otherwise stated. Personal is assumed at a fixed 250.000 € per year. Electricity prices, feedstock prices and CO_2 certificate prices are given in table 3.1.

Parameter	unit	Value	Range
CO_2 emission	$\text{€}\,T^{-1}$	100	0...200
CO_2 feed	$\text{€}\,T^{-1}$	100	200...0
Electricity price	$\text{€}\,MWh^{-1}$	60	50...200
Waste plastics	$\text{€}\,T^{-1}_{CH2-equivalent}$	100	-100...200
Deionized water	$\text{€}\,T^{-1}$	2	-
Ash disposal	$\text{€}\,T^{-1}$	200	-
Natural Gas	$\text{€}\,T^{-1}$	500	300...1100

Table 3.1: Costs of electricity, feedstock prices and emission certificates in the base case along with parameter range used in the feasibility calculation, values based on [97–99].

CO_2 as a feedstock chemical is assumed to cost 200 € per ton, which is attributed to extraction from flue gas, from which the cost for avoided emissions must be subtracted. The yearly costs are composed of CapEx and OpEx. The production costs are calculated based on 330 days per year operation corresponding to 7920 hours. The yearly production thus is $33,000\,t$ of syngas. Production costs are the ratio of CapEx and OpEx, including personal costs, feedstock, CO_2 certificates and maintenance, and yearly production. The mass balance of each process and the utilized equations to determine them are shown along with the calculated results in chapter 7.

Chapter 4

Characteristics of the Reactor, Driver and Discharge

In this chapter, the general properties of the designed reactor are discussed as far as they could be determined. This includes the reactor as well as the discharge itself. As mentioned in the experimental section, the tools used to characterize the discharge were limited. However, basic properties could still be determined. These give insights into the reasons for the performance of the reactor in the splitting reaction of CO_2 and how it could be improved. First, results concerning the discharge structure are discussed in section 4.1. The results obtained show that the plasma operates in the glow-to-arc transition region. Then the drivers capabilities are presented in section 4.2. Here, capabilities and limitations of the driver and the implications for scaled application are discussed. The behaviour and output characteristics under load are illustrated. The use of a magnetic field to stabilize the discharge plays a key role and it is investigated in section 4.4. Lastly, the energy efficiency of the driver is studied in section 4.5. An analysis of the loss mechanisms gives clues on how to further scale the reactor efficiently.

4.1 Discharge Structure

The discharge was first studied in a partially open configuration using the mk 1 reactor under CO_2 flow. This was done to be able to insert a probe into the reactor. The air contamination results in a purple color of the discharge due to light emission from nitrogen. Fig. 4.1 a) shows the discharge in a cone and d) in the plane configuration. The blue color observed on both electrodes is possibly due to sputtering of the steel electrodes, which results in blue light emission from iron ions. The images were taken at slow shutter speed, so it is not possible to determine if the plasma is homogeneous or consists of a rapidly rotating filament.

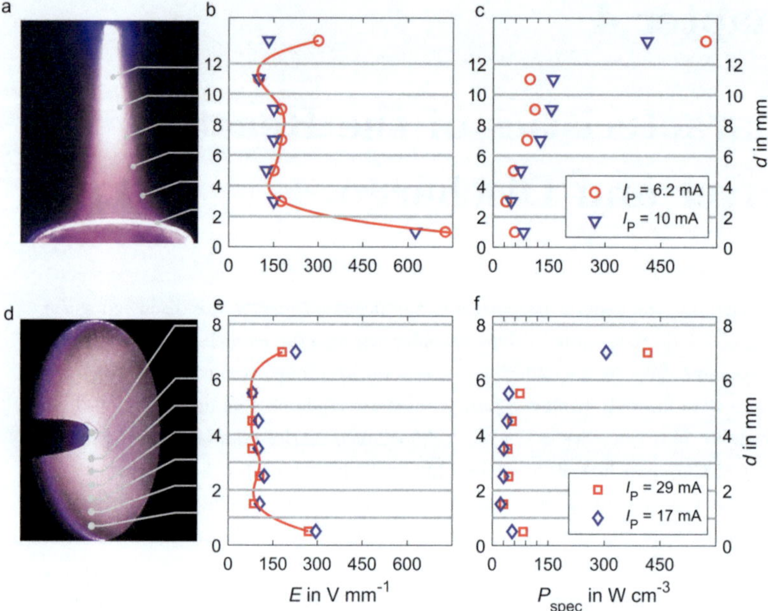

Figure 4.1: View of the discharge generated by the mk 1 reactor. The electrodes are in a cone configuration in a), while they form a plane in d). Grey lines indicate the wire probe insertion points. b) and e) show the respective calculated electric field. A interpolation is given as a line. c) and f) illustrate the power density under the assumption that power is distributed homogeneously in each segment [94].

4.1.1 Discharge Classification

The distinction between glow and arc discharge is a controversial topic since features of both can be present in the glow-to-arc transition region. The electric field within the discharge was calculated and interpolated in the mk 1 reactor. It is shown in fig.

4.1. It can be compared to the electric field observed in an ideal glow- or arc discharge as described in section 2.1: The cathode drop voltage is clearly visible in both electrode configurations in fig. 4.1 b) at a distance of $d = 0...2\,mm$ and in fig. 4.1 e) also at a distance of $d = 0...1\,mm$. The following region between $2\,mm < d < 4\,mm$ has lower luminosity, at least in a), as well as a lower electric field. It can be interpreted as cathode dark space. The electric field is consistently decreasing with applied current, which is typical for plasma in the glow-to-arc transition region. For these reasons, the discharges employed in this work are referred to simply as DC-discharges, although the term 'glow-to-arc-transition-discharge' might be more adequate. To correctly identify the glow-to-arc transition region, it is sufficient to measure a negative differential resistance [48]. Since the wire probe is difficult to include in an enclosed reactor and not much rigidity lies in the distinction between the two types of discharge, the measurements were not repeated for the mk 2 reactor. The voltage-current characteristics still indicate a glow-to-arc-transition discharge there, which is shown in fig. 4.9.

In the mk 2 reactor, the discharge was studied in a pure CO_2 atmosphere. The light emitted from a CO_2 plasma is white or blue, depending on the temperature. At high temperatures, the emitted light is white. At low temperatures, a blue light emission can be seen that is due to oxygen-carbon monoxide recombination [100] - the blue light emission can be distinguished from light emission by metal ions, because in the mk 2 reactor copper electrodes are used. The images of the discharge given in fig. 4.2 show mostly white to blue light emission.

In contrast to mk 1, the magnetic field used in mk 2 is controllable. This enables the acquisition of images at varying magnetic flux B. In fig. 4.2 c), the discharge can be seen without a magnetic field. Thermal currents in the gas carry the discharge up the pin electrode. A moderate magnetic field of $16\,mT$ is enough to force the discharge into the electrode plane, which can be seen in d). It rotates slowly, leaving a visible tail. Increasing the magnetic field further gives the discharge a homogeneous look, which can be seen in a). In b), a $1/4000\,s$ exposure image is shown, where a single discharge channel is visible. In this channel, current density, temperature as well as light emission are highest. The rest of the discharge plane is filled with a visible after glow. The filament forms a straight line, although it could be expected to form a spiral due to the Lorentz force, since a strong magnetic field is present. This is further discussed in section 4.4.

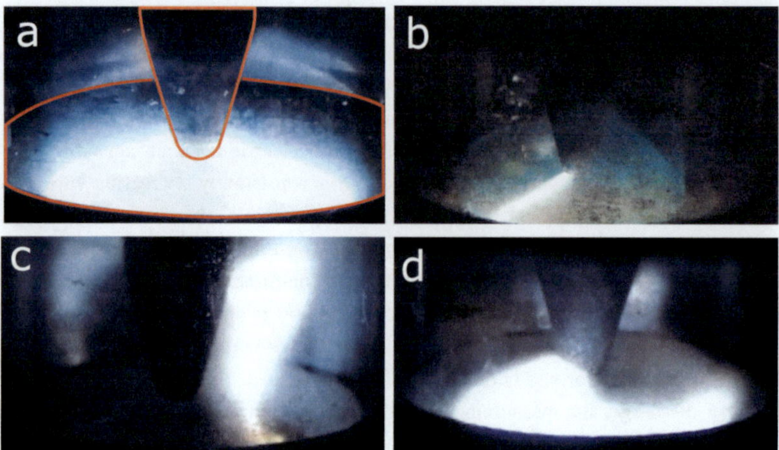

Figure 4.2: View of the discharge generated by the mk2 reactor during operation.
a) Shows an image with regular exposure time, the discharge appears as a disk.
An image at 1/4000s exposure reveals a discharge channel in b). Image c) depicts
operation without magnetic field. In d) a weak field causes the discharge to leave a
visible tail [95].

4.2 Ignition and Operating Range

The plasma can be ignited by the utilized driver circuits in all instances without
issues. In mk 1, the discharge is ignited and sustained by the multiplier circuit. This
makes the driver construction easier, but limits the operation current.The ignition
voltage is consistently slightly lower than the ideal breakdown voltage of CO_2, which
is in the range of $3000\,V\,mm^{-1}$ [79]. Mk 2 ignites at $25\,kV$ corresponding to approximately $2.5\,kV\,mm^{-1}$; ignition voltage increases slightly when a higher magnetic
field is applied, although the voltage increase is inconsistent. In mk 2, the discharge
is ignited by the multiplier circuit but sustained by a single diode rectifier. Thus,
more current can be drawn. This manifests in a larger operation range.

If too little power is applied, the discharge becomes unstable. The current through
the discharge fluctuates, and if it becomes too low, the discharge is extinguished,
which can be seen in fig. 4.3. It reignites shortly after, but since ignition voltage is
higher than burn voltage, a delay appears while the applied voltage increases. The
voltage necessary for re-ignition is lower than for the initial ignition, since the gas in

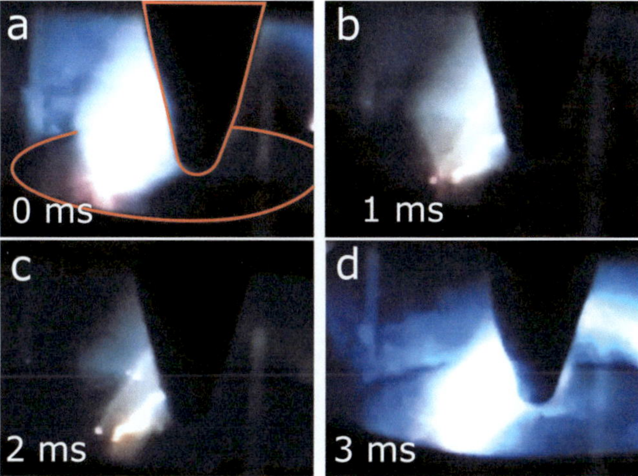

Figure 4.3: Discharge in the mk 2 reactor in operation at low power. The images a-d were acquired as individual frames from slow motion video. They show a discharge channel extinguish and then reignite in a nearby location in quick succession. The electrodes are highlighted in a) [95].

the discharge gap is still partly ionized. In effect, single sparks are being generated, rather than a sustained discharge.

The maximum possible discharge current is limited, especially in the mk 1 driver circuit, by the capabilities of the HV diodes. As described before, the discharge displays a negative V/I-curve. The negative differential resistance leads to a decreasing burn voltage as discharge current increases. This effect limits the power that can be introduced into the plasma. At high applied power (up to $350\,W$ in mk 2), the discharge voltage drops to just $600\,V$. The discharge current is close to $0.6\,A$, the maximum rating for the utilized diodes. To increase the applied power further, a larger electrode gap or higher magnetic field would be necessary.

4.3   Voltage and Current Characteristics

The negative differential resistance can be observed in fig. 4.4. The driver output is indicated as grey lines. The negative differential resistance can be observed.

The output curve of the driver intersects the discharges' VI-curve while falling at a steeper rate, thus satisfying the condition set in section 2.3 for stable operation. The VI-curves in the mk 2 reactor show a similar behaviour, they are discussed in the following section 4.4.

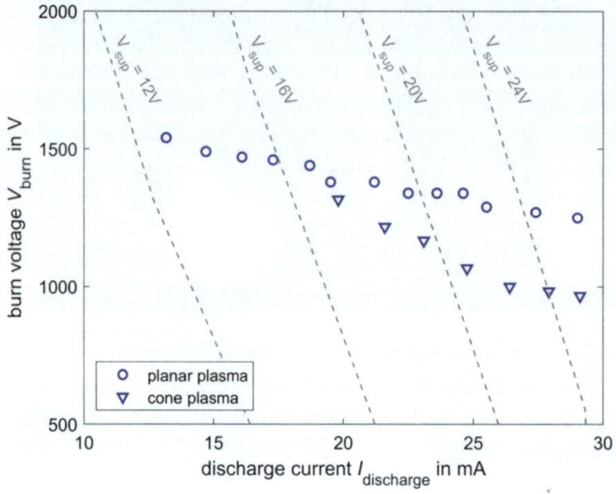

Figure 4.4: Voltage-current characteristics of the mk 1 driver circuit. Grey lines indicate the output of the driver at varying load resistance and constant input voltage. Marks indicate operation points with an ignited plasma. The difference between multiplier stages in each experiment was considered in data evaluation [94].

In fig. 4.5 a), a time-series of voltage and current of the mk 1 reactor is shown. The voltage ripple is high - however, current never drops to zero. The period of the observed fluctuations of $T = 15\,\mu s$ matches the drivers oscillation period. Plotting burn voltage V_{burn} over discharge current $I_{discharge}$ in fig. 4.5 b) results in an oval shape with a sharp edge appearing a high current and low voltage. A similar result was found by Watanabe et al. [101] as well as Dyatko et al. [102], who studied glow-to-arc-transition in argon. The hysteresis that can be seen is attributed to transition between discharge modes: At decreasing voltage, glow-to-arc transition occurs. An arc is formed at high current. It is sustained, while the low resistance drains the the output capacitor of the driver circuit. When current decreases enough, voltage increases which is a signal of arc-to-glow transition and is accompanied by increasing burn voltage. The applied current at which this transition appears is similar to literature [101, 102], although the Lissajous plot takes a more oval shape. This is attributed to the reactive current which is measured superimposed on the discharge current. In mk 2, the ripple voltage is lower, depending on the operation point as low as 15%. No evidence for the glow-arc-transition could be identified in the mk 2 reactor. This is consistent with the higher discharge current, which is above the threshold observed in the mk 1 reactor.

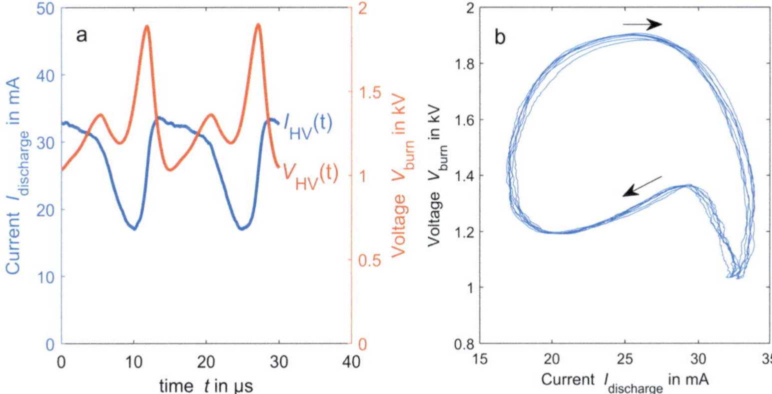

Figure 4.5: Burn voltage V_{burn} and discharge current $I_{discharge}$ in the mk 1 reactor. In a), a time series of both values is shown. b) shows a Lissajous plot of both time series [94].

4.4 Influence of the Magnetic Field

The magnetic field was first introduced to increase the gas volume swept by the discharge. This works so well in the mk 1 reactor that the discharge appears to the naked eye as a disk instead of a filament. More detailed insights into the discharge geometry could not be established at this point. This question lead to the requirements for the next generation reactor. The mk 2 reactor allows tuning of the magnetic field in order to quantitatively study its effects. A key question is the necessary magnetic flux density B to achieve a set rotation rate of the discharge channel and thus full sweeping of the working gas. Indeed, experiments using the mk

2 reactor and low flux density produce a very slowly rotating channel that leaves a bright tail and can be observed with the naked eye. An image is given in fig. 4.2 d). The discharge channel remains nearly straight even at high flux density $B = 32\,mT$. This can be seen in image 4.2 b).

The electric field in this operation point is in the order of magnitude of $E >$ $100\,Vmm^{-1}$. This exerts a force F_E on an electron that can be calculated to be approximately $F_E = 1.6 \cdot 10^{-14}\,N$. The Lorentz force F_B on the other hand can be calculated under the assumption that the electrons acquire a kinetic energy of $E_{el} = 1\,eV$ which is typical for gliding arcs [43]:

$$F_B = qv \times B = 1.602 \cdot 10^{-19}\,C \cdot 6 \cdot 10^5\,m\,s^{-1} \cdot 0.1\,T = 1 \cdot 10^{-14}\,N \tag{4.1}$$

Both forces are similar in magnitude at a magnetic field of $0.1\,T$. However, the amplitude of F_B is also dependent on electron velocity. The utilized magnetic field is below $B < 0.05\,T$. This could be one reason why the discharge channel appears straight: The effect of electrostatic force dominates in this range of operation. This is in opposition to gliding arc discharges that utilize thermal arcs at high current and low voltage, such as [103, 104]. Since the electric field reported in literature arc discharges is lower, the force F_E is also smaller. Thus, the channel can lengthen in response to external factors like gas flow or magnetic fields. Examples are given in fig. 4.6. In a), a very high current and low voltage are applied to generate a diffusive discharge that reach a fast and stable rotation. In b), the discharge channel elongates before reigniting in a straight line. In c), only the cathode end of the discharge channel reconnects. In d), a stable rotation is reached where both electrode attachment points move homogeneously. Similar experimental conditions in c) and d) lead to similar results as seen in the mk 2 reactor.

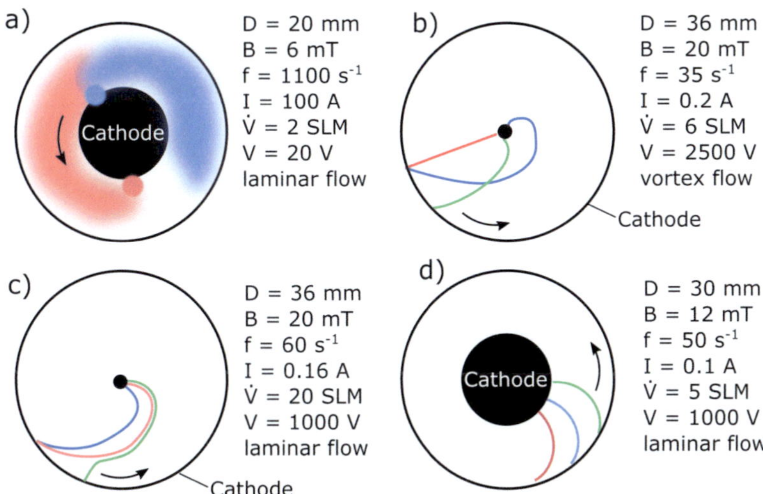

Figure 4.6: Discharge channels observed in literature experiments alongside experimental conditions: Reactor diameter D, applied magnetic field B, rotation frequency f, applied current I, gas flow $\dot{V}$, and burn voltage V. Blue, red and green show time progression. a) illustrates a diffuse gliding arc [104]. b) shows a long channel that does not reach stable rotation but instead reconnects [105] after elongating. Reconnection can also only appear at the cathode, as shown in c) [106]. According to Gangoli et al. [107], a stable rotation can be reached where both electrode connection points move smoothly, which is illustrated in d).

The period of rotation T of the discharge channel depends on the flux density B and discharge current $I_{discharge}$. The period of rotation and shape of the discharge are studied using high speed video. Additionally, the burn voltage V_{burn} is measured using an offset pin electrode as described in chapter 3. This relation between flux density and period of voltage fluctuation is shown in fig. 4.7. It can be seen, that discharge current $I_{discharge}$ and flux density B both correlate with lower period of voltage fluctuation as expected.

The used method of measuring period of rotation by placing the pin electrode off centre is limited. A major problem is, that the voltage observed shows two overlaying signals. A change in discharge voltage due to the change in electrode distance is

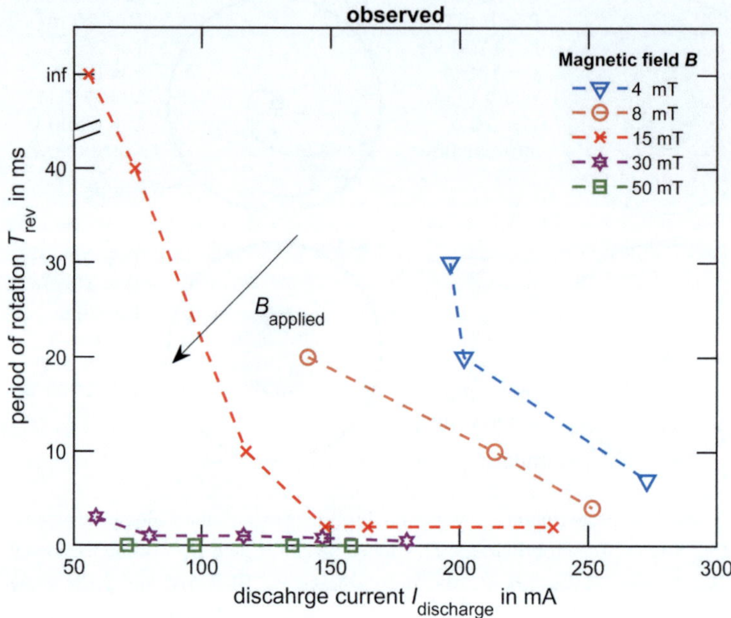

Figure 4.7: Observed period of rotation of the discharge channel at varying current I and magnetic field B measured by analysis of the burn voltage V_{burn} using an off-centre pin electrode [95].

the desired observation. This can indeed be observed. However, a second signal is overlaid: An exponential increase in discharge voltage with small amplitude that is terminated by a sudden drop. This can be interpreted similarly to observations made by Zhang et al. [105]: Near the centre of the circular electrode setup, the circumference is shorter than along the outside. However, the discharge channel reaches a constant velocity over its whole length at any given applied current and magnetic flux. Thus, the inner parts of the channel reach higher angular velocity ω, since $\omega = v/C$ with circumference C. The channel is deformed into a spiral and lengthened, thus discharge voltage increases. After a short period that depends on the discharge current and flux density, the channel jumps back to a straight line, causing a sudden drop in burn voltage. Images of the discharge were analyzed and compared to the fluctuation in burn voltage. Both suggest, that the discharge channel never deviates much from a straight geometry: Voltage ripple was less than 15% and images depict a straight discharge channel in most instances.

4.4.1 Ideal and Observed Rotation Rate

The period of rotation is limited by the drag force on the discharge channel. The channel is thus treated as a 'wire' with a fixed diameter and drag coefficient [107], as described in section 2.1.

Assuming a constant channel diameter d, the period of rotation at a given current can be calculated from the condition $F_{drag} = F_L$ [107]. The Lorentz force acting on the entire channel is simply

$$F_L = l_{channel}\, I_{discharge}\, B\,.\tag{4.2}$$

The drag force over the length of the channel is

$$F_D = \int v(l)^2\, c_p\, \rho\, d_{channel}\, dl = \frac{1}{\sqrt{2}} v_{max}^2\, c_p\, \rho\, d_{channel}\, l_{channel}\,.\tag{4.3}$$

with $v_{max} = 2\pi\, f\, D$. This gives an estimate for the velocity $v_{channel}$ as

$$v = \sqrt{\frac{I\,B}{c_p\,\rho\,d}}\tag{4.4}$$

and the period of rotation as

$$T_{rot} = \frac{\pi D}{v}.\tag{4.5}$$

This calculation neglects the increase in channel length during operation. Also, the change in gas density has to be considered. At a temperature of $\vartheta = 1000\,K$ of the neutral gas and varying gas composition, the density drops by up to a factor of 5 compared to standard conditions. The density is assumed to be proportional to applied power at roughly $\vartheta \sim 3\,K\,W^{-1}\,P$ for simplicity. The channel diameter is assumed to be approximately $D_{channel} = 1\,mm$. Using these assumptions, the periods of rotation shown in fig. 4.8 can be calculated.

The difference in the rotation period observed in figure 4.7 and calculated period of rotation shows the limits of the used measurement method. Furthermore regarding the results found in section 4.4, the short measured period of rotation can be attributed to the jumping of the discharge channel. This means that the voltage fluctuations measured in fig. 4.7 must be attributed to reconnections of the discharge channel and not, as expected, rotation rate. A fast period of 'jumps' or

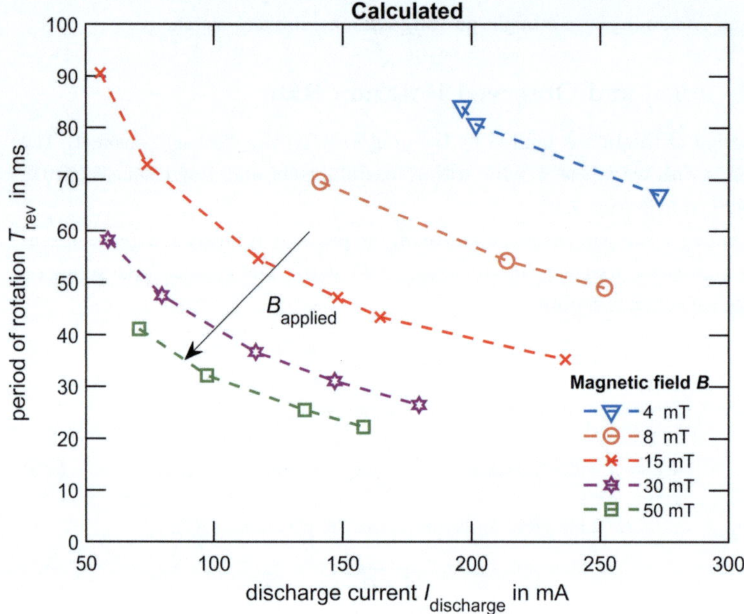

Figure 4.8: Calculated period of rotation of the discharge channel at varying discharge current $I_{discharge}$ and magnetic field B under the stated assumptions.

reconnections suggests, that the discharge channel cannot deviate much from the straight geometry at which it is ignited. Reconnections happen at a higher rate than reported in literature. E.g. Zhang et al. [105] measured a rate of jumping roughly three times as high as the period of rotation, corresponding to roughly 200 Hz. Meanwhile, Wang et al. [106] concluded that the reconnections occur at a rate of three times the rotation rate. A likely reason for the difference is the driver circuit: Gangloli et al. [107] suggest that the discharge forms a new channel when the power available for the discharge given the driver output characteristics drops. Since most studies use a ballast resistor, this occurs when the resistance of the discharge channel $R_{channel}$ and ballast $R_{ballast}$ are equal. This is not the case in the resonant power supplies constructed in the scope of this work, which could explain the high frequency of voltage fluctuations measured. A second reason for the varying reconnection frequency could be the ratio of cathode and anode diameter, which determines the relative amplitude of velocity of the discharge channel of both electrode contact points.

The criterion for reconnection of the discharge channel can be used in the design for future driver circuits: Assuming constant current, the plasma power will fluctuate in a predictable pattern between the ignition and maximum elongation.

4.4.2 V/I Curves in Relation to Magnetic Field

The rotation of the discharge channel has two major effects: First, fresh gas is fed into the discharge zone at a higher rate, which in turn decreases the channel temperature. As a result, the particle density in the channel is higher than in a case without magnetic field. The resistivity of the channel is thus larger, which leads to a larger burn voltage when the magnetic flux is increased. Second, the filament has a greater average length. Thus, a higher burn voltage is necessary to support it, even at similar resistivity per length. The influence of both effects could not be quantified, since neither could be directly measured. However, the result of both effects can be seen in fig. 4.9: The V/I-curve of the discharge flattens and burn voltage V_{burn} increases. Usually, the mean free path of electrons and the degree of ionization both increase with applied power. Both quantities decrease resistivity. Since more fresh gas is fed into the discharge channel when a magnetic field is present, resistivity remains high at increased applied power. Thus, the discharge is stabilized. The now flatter V/I-curve reduces the requirement for the current limiting capabilities of the driver circuit. Furthermore, a higher discharge power is possible in a similarly sized reactor. The results are similar to experiments performed by Wang et al., who were studying the synthesis of carbon nanoparticles by plasma [106].

Electric field E depends on discharge current I and magnetic field B. The maximum discharge power in a reactor with electrode distance D and strain of the discharge channel ϵ can be calculated as

$$P = E(I, B) \, I \, D \, \epsilon. \tag{4.6}$$

Strain ϵ describes the length ratio of the discharge channel to the electrode distance. At a maximum discharge current of $I_{discharge} = 300\,mA$ and the gap of the mk 2 reactor of $d = 10\,mm$, the maximum power is $P = 195\,W$.

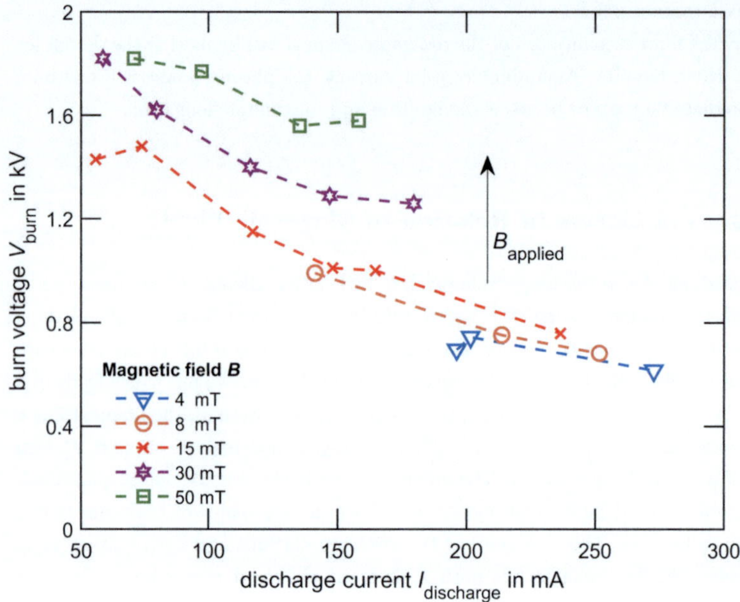

Figure 4.9: Measured burn voltage of the mk 2 plasma reactor at varying applied power and magnetic field [95].

When a magnetic field is present, the maximum power, calculated by extrapolation of the measured data, is $390\,W$ at $B = 30\,mT$. The scaling effects are a key parameter regarding the applicability of plasma systems. While in microwave plasmatrons, discharge power P scales with the cube of the diameter of the reactor, the same is not likely for direct current atmospheric pressure discharges. A scaling law proportional to $P \sim B\,D^{1...2}$ is expected, since discharge voltage grows linearly with diameter and current can be either constant or scaled linearly to achieve constant power density. The determination of scaling laws must be subject of future studies. A key question is the maximum discharge channel strain ϵ.

4.5 Driver Efficiency

To increase the overall energy efficiency of the process, the efficiency of the driver circuit must also be high. A realistic goal for energy efficiency is 90%, including ignition circuit. This target efficiency will be justified in the following sections. Both drivers did not achieve this goal. However, the mk 2 driver is theoretically able to achieve it since loss mechanisms are known. They will be discussed in the following section. Losses were in part measured and in part calculated. The formulas used to estimate losses for components are found in section 2.3. The mk 1 driver circuit reaches a maximum energy efficiency of $\eta_{max} = 79\%$. The losses are illustrated in fig. 4.10.

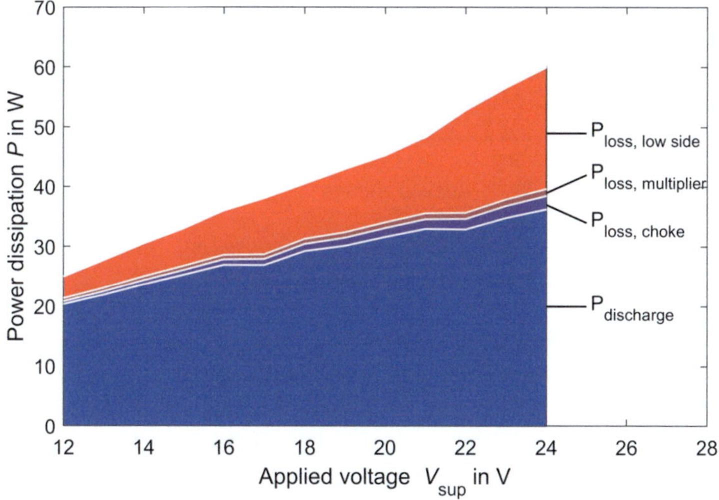

Figure 4.10: Power dissipation in the mk 1 driver circuit at different applied voltages V_{sup}. Supplied power is the upper outline of the patched area.

They can mainly be attributed to the flyback transformer and MOSFET driver resistor network, summarized in $P_{loss,lowside}$. The losses are calculated as the difference between applied power and discharge power $P_{discharge}$. Losses in the choke inductor L_{choke}, $P_{loss,choke}$ and forward losses in the voltage multiplier $P_{loss,multiplier}$ are calculated using the discharge current. The remaining losses are attributed to the transformer and primary side electronics. In the mk 2 driver circuit, most of the losses can be attributed to the transformer core. They are illustrated in fig. 4.11.

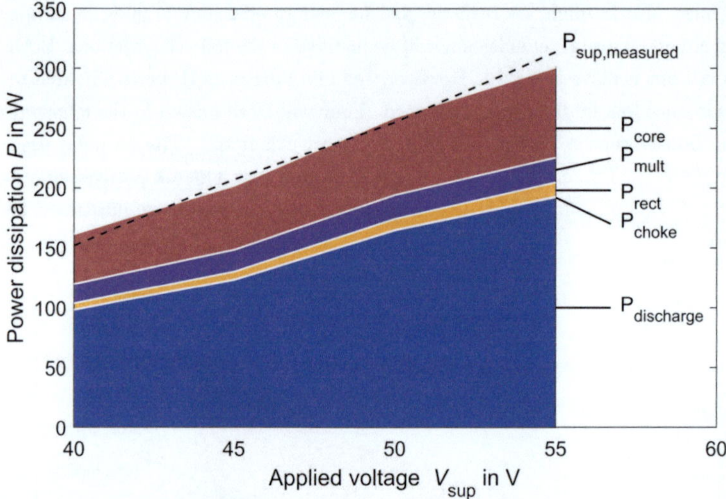

Figure 4.11: Power dissipation in the mk 2 driver circuit at different applied voltages. V_{sup}. Supplied power is indicated as a dotted line.

The core was deliberately chosen to be oversized in order to achieve better insulation for the high voltage side. The core chosen for the custom transformer was a UU-core from *Epkos/TDK* (core with two U-shaped segments) with measurements 126 x 91 x 20 mm. The winding number is kept to a minimum of $n_1 = 6$ and $n_2 = 196$ for simple manufacturing. The oversized core combined with small winding number leads to high magnetization and thus high core losses P_{core}. A properly designed compact ferrite transformer with epoxy insulation can reach efficiencies in excess of 98%. The losses in the voltage multiplier are also significant. This is due to the fact, that a single diode rectifier was used. The amplitude of the transformers secondary voltage in this configuration is approximately

$$V_{secondary} = \frac{1}{2} \cdot V_{sup} \cdot \pi \cdot \frac{n_2}{n_1} + V_{discharge}. \tag{4.7}$$

By using a full bridge rectifier the voltage applied to the multiplier would drop after ignition to

$$V_{secondary} = 2 \cdot V_{discharge}. \tag{4.8}$$

This would drastically reduce the power dissipated in the voltage multiplier after the plasma is ignited. The rectifier uses diodes with a large forward voltage drop and thus contributes with P_{rect} to the lost power. It can be seen that the sum of the calculated power losses and discharge power $P_{discharge}$ approximately corresponds to the applied power. Very little power is lost on the low voltage side. The MOSFETs are switched in resonance, so in the zero-voltage and zero-current state. This is an advantage of the Royer-converter. However, it also has two major drawbacks: First, the resonance leads to a large circulating current in the primary of the transformer. This in turn leads to copper losses, which would be significant at higher discharge power. Second, the MOSFET switches need to be rated for π times the supply voltage due to resonance.

The overall efficiency of the mk 2 driver circuit is in the range of $\eta = 60...65\%$. The calculation gives clues to possible improvements and maximum achievable energy efficiency. By using a full bridge rectifier, the losses in the voltage multiplier could be reduced from $P_{mult}(55\,V) = 20.6\,W$ to just $P'_{mult}(55\,V) = 2.81\,W$. The losses in an adequately sized transformer are in the range of 2%. These changes would increase the overall energy efficiency of the driver to 91%.
These considerations can be summarized in a list of proposed changes to the next iteration of the driver circuit:

- A full bridge rectifier should be used for high voltage rectification.

- A properly sized transformer can help reduce losses. To achieve sufficient high voltage insulation in a smaller form factor, the secondary winding must be coated using epoxy resin.

- The rectifier should be constructed using diodes with lower forward voltage drop.

- An LLC-configuration driven above its resonance frequency leads to lower current in the primary winding, while MOSFETs with lower voltage rating can be used.

- An alternative ignition circuit that uses less power should be implemented.

An LLC-converter including switching signal generator, full bridge converter, custom transformer and rectifier was designed, simulated, built and tested in the scope of this work. Sadly, it could not be finished in time to be tested on a plasma reactor. However, the current limiting capabilities and energy efficiency are very promising. Preliminary results obtained with the designed LLC-converter are shown in appendix D.

4.6   Summary

In this section, the character of the discharges generated with each version of the reactor and driver was discussed. Both versions operate in the glow-to-arc transition region. A varying degree of cathode fall voltage is observed, while the electric field is in the order of $E = 100\,V\,mm^{-1}$. The discharge can be forced to rotate using a magnetic field. At sufficient flux density, a disk-like shape emerges. While the discharge channel seems to remain almost straight while rotating, the resistivity of the discharge changes. This effect is more pronounced at higher magnetic flux B and applied power $P_{discharge}$. The rate of rotation is difficult to distinguish from the rate of reconnection, i.e. the rate at which the cathode spot jumps forward in discrete steps. This rate is likely higher than in literature experiments, which is accounted for by the driver circuit: Reconnection appears, when the discharges operation point passes the point of maximum power. This point is determined by the characteristics of the driver circuit. A brief comparison shows, that the reactor diameter also has a strong influence on rotation rate, as well as placement of the anode and cathode. The shape of arcs obtained is broadly similar to literature experiments. Both drivers were evaluated for their performance. While mk 1 could drive discharges with up to $P_{discharge} = 35\,W$, mk 2 reaches $P_{discharge} = 350W$. Losses in both driver circuits were identified. Notable losses occur mainly in the transformer and voltage multiplier. Further scaling requires a change to a new topology. An LLC-converter is proposed for this application. Preliminary results obtained with an LLC-converter are shown in appendix D.

Chapter 5

CO_2 Splitting in the DC Discharge

The now characterized reactors are put to use in the CO_2 splitting reaction within the experiments described in chapter 3. In the experiments performed with the mk 1 reactor proving general feasibility was the key concern. After this validation, the mk 2 reactor was built to improve on those results and evaluate measures for further performance enhancement. The results are then compared first to similar plasma reactors, then to a broader selection of plasma technologies that are frequently used, and lastly to competing approaches for electrification within the chemical industry.

5.1 CO_2 Splitting in the Mk 1 Reactor

Fig. 5.1 shows the CO_2 conversion and energy efficiency obtained in the mk 1 reactor. The results show an increase in conversion with increasing specific energy E_{spec}, while energy efficiency drops linearly over the same interval. Maximum energy efficiency is $\eta = 29\%$ at a flow of $\dot{V} = 190\,SCCM$ while maximum conversion is $X = 20\%$ at a flow of $\dot{V} = 80\,SCCM$. Energy efficiency and conversion behave contrary to each other. This is expected for non-thermal plasma and has been observed in literature [64]. It was shown in fig. 2.10 in section 2.2.1 that in thermal processes, energy efficiency increases with conversion up to a value of $\eta = 45\%$. It can thus be concluded that in this reactor, the excitation of CO_2 molecules is not occurring through thermally distributed energy states. At the same time, the electric field that was previously measured in the mk 1 reactor leads to an expected average electron energy in the range of $1\,eV$. Thus, the splitting of CO_2 would occur by vibrational excitation [64]. Cold plasma at atmospheric pressure, e.g. in DBDs,

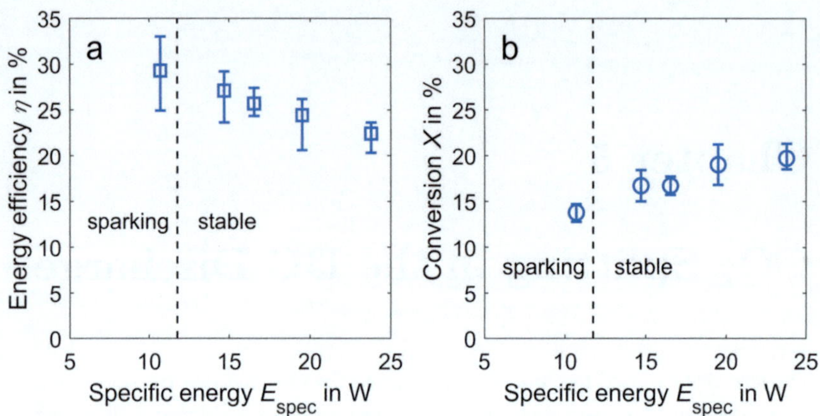

Figure 5.1: Energy efficiency η (a) and conversion X (b) achieved in the mk 1 reactor, measured at varying specific input energy E_{spec} and constant gas flow of $V_{in} = 190\,SCCM$ [94].

was not observed to achieve greater energy efficiency than 20% with the exception of Ozkan et al. who reached 23% using a pulsed DBD [72]. Non-thermal warm plasma is frequently reported to reach higher energy efficiency. Overall, the results show that the mk 1 reactor and pin-to-ring discharges using a magnetic field in general, are well suited for the splitting of CO_2. A potential improvement is the use of a laminar gas flow instead of a turbulent one, which would lead to a more even energy distribution in the working gas.

5.2 CO_2 Splitting in the mk2 Reactor

Fig. 5.2 shows the CO_2 conversion obtained in the mk 2 reactor at varying power and gas flow rate. The maximum conversion reached is $X_{max} = 28\%$ at the lowest measured gas flow of $\dot{V}_{in} = 0.65\,SLM$. At $\dot{V}_{in} = 1.25\,SLM$, a conversion of $X = 27\%$ is still possible.

Like in the mk 1, conversion is increasing with specific energy. However, it does not do so indefinitely. At an applied power of $P_{discharge} = 192\,W$ and gas flow up to $V_{in} = 1.0\,SLM$, the conversion is lower than at $P_{discharge} = 165\,W$. The most

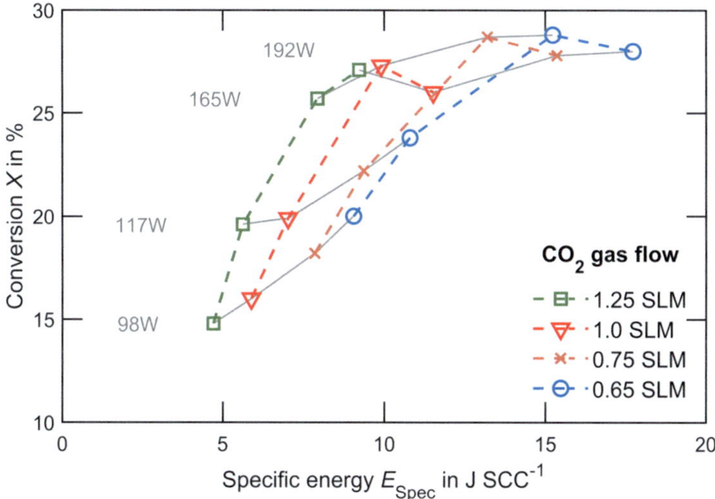

Figure 5.2: Conversion X achieved in the mk 2 reactor, measured at varying power P and varying gas flow $\dot{V}_{in}$. The grey lines indicate constant power, while the dotted lines indicate constant gas flow.

likely reason for this is heating of the reactor, leading to a higher back reaction rate. At this power, the packed bed is glowing in a bright red colour during operation, corresponding to a temperature of above $\vartheta_{bed} = 800°\,C$.

The increase of conversion with applied power is in line with the measured rotation rates of the discharge channel described in chapter 4. As shown in chapter 2, the rate of rotation of a discharge channel is roughly proportional to the square root of the discharge current. Since higher rotation rates lead to better sweeping of the gas, the power is introduced into the working gas more evenly.

The energy efficiency calculated from the same data is shown in fig. 5.3. Plotting all the data, it is apparent that the energy efficiency is mainly dependent on the gas flow rate, while the applied power has a minor influence. The highest energy efficiency that was achieved is $\eta = 44\%$ at $E_{spec} = 117\,W$ and a gas flow of $\dot{V}_{in} = 1.25\,SLM$.

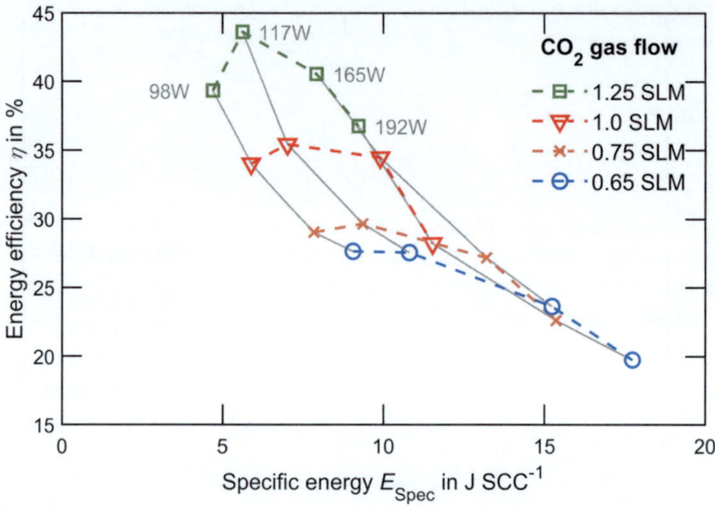

Figure 5.3: Energy efficiency η achieved in the mk 2 reactor, measured at varying power P and varying gas flow $\dot{V}_{in}$. The grey lines indicate constant power, while the dotted lines indicate constant gas flow.

To understand which design factors account for the obtained results, the reactor setup is varied. A variation of the discharge current ripple can be obtained by changing the choke inductance between $L_1 = 1.5\,H$ and $L_2 = 3.0\,H$. No significant effect was observed, although an effect might occur at lower inductance and respectively higher current ripple, which could not be tested. Inverting the direction of gas flow drastically reduced conversion from 20.5% to 10.5%. This could be due to less uniform gas flow, due to the lack of quenching that would otherwise be provided by the packed bed, or due a catalytic effect of the bed. Removing the packed bed resulted in an even more extreme decrease in conversion to just $X = 7.5\%$. Operating the reactor at a weaker magnetic flux also lead to drastically lower conversion. However, the results could not be reproduced consistently and are thus not given quantitatively. A broader overview of relevant factors for performance can be achieved by a literature review.

5.3 Comparison to Literature Data and Review

A variety of similar reactors for the purpose of CO_2 splitting are reported in literature. Mostly, they are referred to as gliding arc reactors regardless of their exact operation regime - thermal and non-thermal plasma reactors are bunched together as well as glow- and arc discharges. Their optimum performance is compared to the results achieved in this work in fig. 5.4. Details about the experimental setup are given in table 5.1.

Source	electrodes	catalyst	quenching	gas flow	magnets	technology
Sun [108]	linear	none	none	laminar	no	GA
Li [109]	linear	none	none	laminar	no	GA
Zhang [110]	linear	TiO_2	on bed	turbulent	no	GA
Ramakers [111]	coaxial	none	none	vortex	no	GA
Li [112]	linear	none	none	laminar	yes	GA
Kim 2020 [66]	N.A.	none	none	vortex	no	MW
this work	coaxial	ZrO_2	on bed	laminar	yes	GA

Table 5.1: Comparison of various reactor setups for CO_2 splitting that demonstrate possible improvements in reactor setup.

It is apparent, that the reactor developed in this work outperforms the ones presented in literature. One possible reason has to be highlighted in advance of in-depth discussion: The volume flow in this study was measured using a manually calibrated rotameter, so some amount of systematic error is expected. For this reason, no record should be claimed without reproduction.

However, when comparing the achieved results to literature, it becomes apparent that multiple advantageous design choices were implemented. None of the GA reactors reported in literature use a packed bed through which the gas flows directly. The packed bed contributes by quenching the gas and potentially also by introducing surface effects. However, no conclusive proof for the benefit of zirconium oxide as catalyst could be identified in literature, e.g. by Uytdenhouwen et al. and Michielsen et al. who studied the use of the material in a DBD reactor [113, 114]. The benefit of quenching is illustrated by Kim et al. [66], who compare a reactor with and without quenching rod and achieve a great improvement in conversion. Zhang et al. employ a catalyst in a packed bed instead of a metal rod to combine catalytic effects and quenching [110]. This led to a roughly 50% increase in energy efficiency and also a 30% increase in conversion. However, the gas cannot flow through the packed bed and the gliding arc is generated between linear electrodes and thus not uniform.

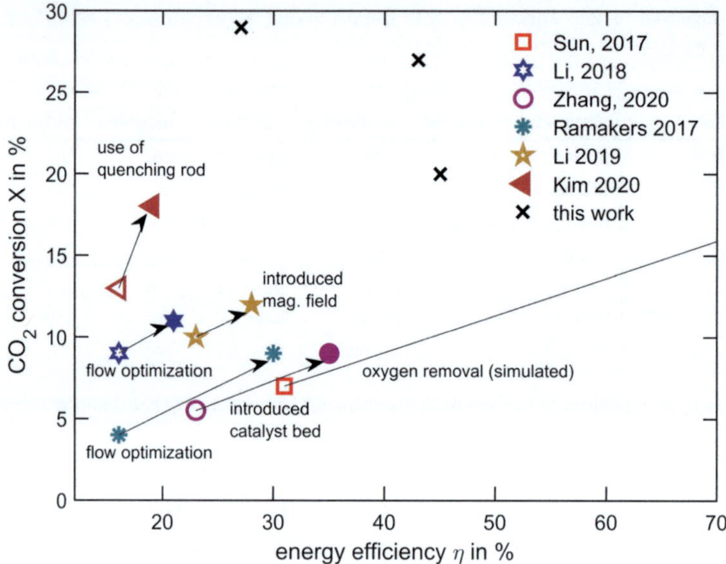

Figure 5.4: Energy efficiency η and CO$_2$ conversion X achieved by different plasma reactors that use similar approaches to design optimization. Symbols without filling indicate baseline results, filled symbols represent results after optimization [66, 108–112].

A key improvement between iterations mk 1 and mk 2 presented in this work is the use of laminar instead of turbulent gas flow. Li et al. [112] and Ramakers et al. [111] both demonstrated great performance improvements by optimization of the gas flow. While Li et al. optimized the reactor shape to obtain a laminar flow instead of a turbulent one, Ramakers et al. tested different coaxial electrode geometries to optimize a vortex gas flow. A magnetic field was used by Li et al. in a followup study to increase the swept volume of a gliding arc with linear electrodes [115]. This is similar to the approach implemented in this work and also improved performance. Lastly, an outlook into more advanced designs is given by Sun et al. [108]. Their simulation shows the potential performance improvement by suppression of the recombination reaction between oxygen and carbon monoxide. While this might not be actionable in practice, it demonstrates the upper limit of performance improvement when an oxygen extractor would be incorporated into the plasma reactor. The performance improvement is literally 'off the chart' in fig. 5.4; an energy efficiency of 86% is calculated in their simulation.

None of the presented references managed to combine all the design choices that have proven advantageous. The closest resemblance to this work is the reactor used by Zhang et al. [110]. The authors presented a coaxial reactor that used an axial magnetic field in a separate paper. However, they had to switch back to a more basic setup with linear electrodes and turbulent gas flow for this study, because the employed ignition mechanism would not support the upside down operation necessary to utilize a packed bed. This shows once more an advantage of the purely electrical ignition used in this work: It gives greater flexibility in the reactor design and allows the incorporation of a packed bed of catalyst.

In summary, various advantageous design choices were made independently and at least partly justify the high performance of the presented reactor.

5.3.1 Comparison to Other Plasma Technologies

Different plasma technologies compete for the highest efficiency in CO_2 splitting. The aim of this section is to identify which processes are suitable for GA reactors and under which circumstances it is reasonable to pursue their development. Besides their technological differences, which were discussed in the introduction in section 2.3.1, they generate plasma with different properties. These are listed in table 5.2. Microwave plasmatrons can generate plasma over a wide variety of parameters and have been reported to reach the highest electron density. Their reduced field and electron density are well suited for driving endothermic reactions. DBDs on the other hand are mainly useful for exothermic or mildly endothermic reactions. This is due to the low neutral gas temperature and low electron density and low electron density in glow mode. In spark mode, a higher density can be achieved. Nevertheless, the main use for DBDs is to drive reactions that have a favorable equilibrium at low temperature, e.g. synthesis of ammonia [116, 117], ozone or methanol [118, 119].

technology	$T_{neutral}$ in K	T_e in eV	E_{red} in Td	n_e in cm^{-3}
MW [120, 121]	$10^3...10^4$	< 1	< 100	$10^{15}...10^{17}$
DBD [64, 122]	$300...1000$	$2...10$	$100...500$	$> 10^{15}$ (spark mode) 10^{12} (glow mode)
Gliding arc [123]	$2...3.5 \cdot 10^3$	$0.4...3.4$	$20...60$	$10^{12}...10^{14}$
Inductive [124]	$0.6...1 \cdot 10^4$	$< 1\,eV$	N.A.	$10^{13}...10^{16}$

Table 5.2: Plasma parameters achieved by different plasma technologies at atmospheric pressure. Note, that only orders of magnitude are given.

They see increasing use for this application recently. Another use is the study of catalysts, since they are easy to incorporate into the discharge volume as packed bed or coating on the dielectric.

Gliding arcs reportedly produce smaller electron densities than microwave plasmatrons. However, they can also achieve low electron energy. The temperature within the discharge channel is given in literature as $T_{neutral} = 2000...35000\,K$. It is questionable, if this is true in all operation parameters. No values could be researched for gliding arcs that utilize a magnetic field or ones that are operated at particularly low discharge current. To explain the observed performance reported in this work, a more detailed characterization of the plasma seems necessary, especially of the influence of magnetic field and current on each species' temperature and density.

Inductively coupled plasmatrons are rarely used at atmospheric pressure. When they are, the reported parameters suggest very high operation temperature, which is not ideal for efficient implementation of chemical processes, but can be used to drive plasma torches. The plasma parameters dictate results that can be achieved in general. It appears, that gliding arcs as well as microwave plasma can have favourable properties. However, once implementation goes beyond lab scale, a new set of properties gains relevance: How well can a technology be integrated into industrial systems? Some key parameters are given in table 5.3. To be industrially viable, a plasma source must be scaleable at least to multiple kW scale in decentral applications. Parallelization is an option but only economic to a certain extent. Microwave plasmatrons in this power range are already commercially available.

technology	size scaling	driver efficiency	wall heat flux	plasma catalyst	post p. catalyst	O_2 removal
MW	$P \sim l^3$	76%	medium	no	yes	yes
DBD	$P \sim A^2$	90%	high	yes	yes	no
GA	$P \sim l^{1...2}$	88%	low - medium	?	yes	yes
Inductive	$P \sim l^3$	87%	medium	?	yes	yes

Table 5.3: Possible process integration and scaling of different plasma technologies. O_2 removal describes the ability to integrate high temperature oxygen ion conductors or other high temperature membranes into the reactor.

Power scales with the cube of reactor size. However, a limitation is the wavelength of the employed waves: A $2.45\,GHz$ -plasmatron produces radiation with a wavelength of $120\,mm$. A homogeneous plasma can only be excited in a volume less than half the wavelength. $915\,MHz$-plasmatrons can be used when scaling the technology, which is viable in principle. In a DBD reactor, the power per area of dielectric is constant when other parameters are kept constant. Stacking of dielectrics easily increases system power. Stacking is hindered by the poor thermal conductivity of the dielectrics which impedes heat transfer out of the reactor. The scaling laws for gliding arcs are still not entirely determined as previously discussed. Likely the power can be scaled either linearly or quadratic to the electrode distance - less easily than in microwave discharges. The power is limited by the discharge voltage that can be reasonably implemented, since changes in discharge current will change its properties. Inductively coupled plasmatrons are in use in research facilities that transform power on the megawatt scale. Here, scaling is seemingly not an issue.

But what about the energy efficiency of a typical driver for each type of plasma? Two ways are available to generate microwave power: At high power, magnetrons are common, while at low power semiconductor devices can be used. In a magnetron, up to 30% of the applied energy are lost at an excitation frequency of $2.45\,GHz$, while lower losses on the order of 15% are possible when $915\,MHz$ are used instead. The difference is due to the kinetic energy of electrons that generate the microwave radiation, which is lost when they inevitably strike the anode. Another 5% of the energy may be lost in the rectifier that is necessary to drive microwave plasmatrons. In addition, up to 5% of energy are not dissipated in the plasma but instead reflected back into the cavity. In order to not destroy the magnetron, the energy is dissipated in a dump load. This gives a maximum driver efficiency of $\eta_{max,MW} = 76\%$ for a typical system [125]. Higher energy efficiency could be achieved by solid state magnetrons that directly convert low DC voltage to microwaves at reported efficiencies of 65% (e.g. Ampleon BLC2425M10LS500PZ). This values is expected to increase in the future. In a DBD plasma, a rectifier and alternator stage that will probably utilize a transformer are necessary. However, a resonant driver can be used which allows for high driver efficiency of up to $\eta_{max,DBD} = 90\%$. To drive a non-thermal gliding arc plasma, the same components are necessary in theory, e.g. in the form of a LLC-converter as described in section 2.3.4. In the case of a DC discharge, an additional rectifier is necessary. When properly built, this reduces efficiency only insignificantly - high voltage diodes with a ratio of blocking voltage to forward voltage of more than 1:900 are commercially available. An energy efficiency of $\eta_{max,GA} = 88\%$ thus seems achievable.

Inductively coupled plasmatrons also require a rectifier and alternator stage, but no transformer. The alternator can be very efficient since it is switched in resonance. Huge currents can occur in the resonant inductance, but nonetheless a coupling efficiency of up to 95% was reported in literature [126] while 92% may be a typically achieved value. Assuming the sum of other losses at 5%, an inductively coupled plasmatron may achieve up to $\eta_{max,ind} = 87\%$.

Depending on the chemical process to be implemented, it can be desirable to introduce additional components into the plasma reactor. A catalyst can be placed on a packing material in the plasma volume or downstream from it. Placing it in the plasma is only possible in DBDs and might be possible in GAs. The dielectric or conductive properties do not allow a plasma to form in the vicinity of the packing material in the case of inductively coupled or microwave discharges. Placing a catalyst downstream from the plasma is possible with all technologies. In the case of CO_2 splitting, the removal of oxygen is a key success criterion. Oxygen removal using selective technologies usually requires high temperature. For this reason, porous membranes or ion conductors can be implemented in Microwave-, Gliding arc-, or inductive plasma reactors, but likely not in DBDs.

In summary, gliding arcs have favourable properties to enable endothermic chemical reactions that achieve high conversion at high temperature, which they share with microwave plasma. They can be generated at a higher efficiency than microwaves and could utilize similar catalysts and reactor geometries. However, questions remain regarding scaleability.

5.3.2 Comparison to Other Fields

Further generalization leads back to the question which technological approach is generally best suited for the realization of an electrically powered chemical industry. The most established branch of technologies is thermochemistry that accounts for the bulk of todays chemical processes. The younger contestants are high- and low temperature electrolysis, as well as photochemistry, biotechnology and plasma technology. They are distinct enough that each will probably find industrial application in the future [13]. However, the same question must be asked as before with the different plasma technologies: How do they compare in performance?

The considerations made in this chapter have previously been published [127]. To compare the performance of each technology in CO_2 splitting, energy inputs must be compared. This can be done on the 'cell'-level: For electrolyzers, applied voltage and current are used to calculate applied power. Additional product cleanup steps are not considered. Conversion is considered. To compare the results to thermochemical reactors, the chemical- and heat energy that is used in those must be included. CO_2 is usually produced by means of water-gas-shift using renewable hydrogen, so this can be done easily enough. Hydrogen is assumed to be made in electrolysis cells at an energy of $350\,kJ\,mol^{-1}$.

The energy used by the plasma systems is assumed to be the only the applied power to the reactor, so conversion and energy efficiency obtained in experiments performed in this work are directly compared. The optimum operation point achieved an efficiency of $\eta = 43\%$ at a conversion of $X = 27\%$. Lower pressure operation can increase energy efficiency of plasma systems. Recently, Huang et al. [128] achieved an energy efficiency of $\eta = 60\%$ at a conversion of 23% using a radiofrequency discharge at reduced pressure of $p = 100\,mbar$. In order to compare the results, compression of the product gas to atmospheric pressure must be included. The energy necessary to do this per mol of CO is approximately

$$W = n\,R\,T\,ln\left(\frac{V_2}{V_1}\right) \rightarrow \frac{W}{n} = 8.314\,kJ\,mol^{-1}K^{-1} \cdot 300\,K \cdot ln(10) \approx 6kJ\,. \quad (5.1)$$

In this equation, n is the amount of substance in mole, R is the ideal gas constant, T is temperature and V is volume. Assuming an energy efficiency of 60% for the compressor still only slightly impacts the process efficiency to $\eta = 59\%$.

The highest reported energy efficiency identified in literature for a low temperature CO_2 electrolysis was observed by Liu et al. [109] who reached a viable operation point at an applied voltage of $V_{cell} = 3.0\,V$ and a coulombic efficiency of $\eta_C = 98\%$.

The resulting energy efficiency can be calculated by

$$\eta_{cell,LT} = \frac{1}{\eta_c} \frac{283\,kJ\,mol^{-1}}{n\,F\,V_{cell}} = 48\%. \tag{5.2}$$

F is the Faraday constant. The conversion achieved by Liu et al. was $X = 24.5\%$ in the same operation point. In the case of high temperature electrolysis, the utilized heat energy must also be included into the calculation. No heat recovery is assumed in this simplified calculation. The most efficient system identified in literature, published by Kaur et al. [129], uses a solid oxide fuel cell at $800°C$. The heat energy can be included in the calculation approximately by

$$\eta_{cell,HT} = \frac{1}{\eta_c} \frac{283\,kJ\,mol^{-1}}{n\,F\,V + \Delta T\,c_p(CO_2)}. \tag{5.3}$$

$c_p(CO_2) = 44\,kJ\,mol^-1$ is the mean heat capacity of CO_2 per mole in the temperature range and ΔT is assumed to be $800\,K$ corresponding approximately to heating from ambient temperature. When the cell is run at just $V_{cell} = 1.5\,V$ as published, an energy efficiency of $\eta_{cell,HT} = 85\%$ can be achieved. Electrochemical stability of the employed materials limits the conversion to just $X = 10\%$. The highest energy efficiency identified in literature using an electrically driven thermochemical reactor was achieved by Zonetti et al. [130]. In the study, an electrically heated RWGS reactor is used to produce carbon monoxide. The used energy is the sum of energy used in hydrogen production and energy used to heat the reactor. Hydrogen is in turn provided by electrolysis which uses approximately $W_{H2-electrolysis} = 350\,kJ\,mol^{-1}$. Heat energy is necessary to heat up CO_2 and hydrogen stoichiometrically, using the mean heat capacities $c_p(H_2) = 30\,kJ\,mol^{-1}$ and $c_p(CO_2) = 43\,kJ\,mol^{-1}$. Since the reactor operates at $700°C$, the energy efficiency is

$$\eta_{RWGS} = \frac{283\,kJ\,mol^{-1}}{W_{H2-electrolysis} + \Delta T(c_p(H_2) + c_p(CO_2))} = 62\%. \tag{5.4}$$

At the same time, a conversion of 48% is achieved. All results are summarized in fig. 5.5. The results clearly reflect the fact that thermochemical systems had a head start in development of multiple decades; they reach the most convincing combination of high energy efficiency and conversion.

Solid oxide electrolysis of CO_2 is an attractive one-step solution. Compared to plasma systems an advantage is, that oxygen is not found in the product gas. However, two major concerns remain: The long term stability in load cycling and the

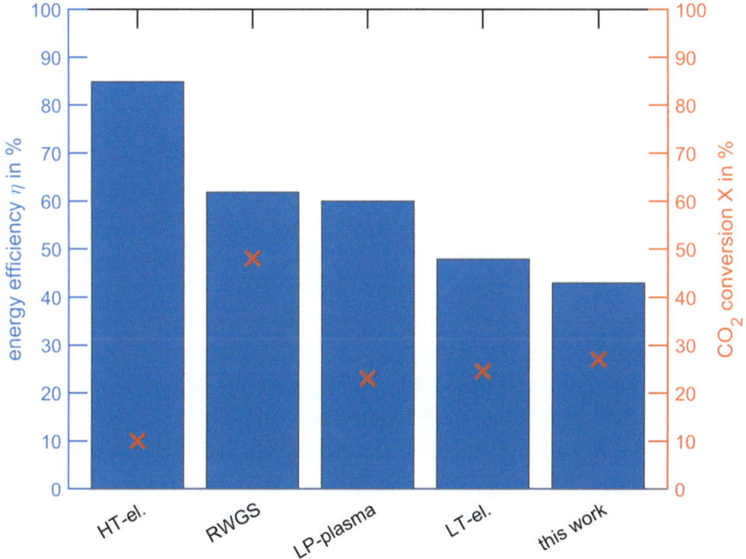

Figure 5.5: Energy efficiency and conversion achieved with various electrical systems for CO_2 conversion. Each entry represents a best-of-class identified by literature review. Adapted from [127].

electrochemical stability of the materials and coking at high conversion rates. Electrolysis of CO_2 seems to have a smaller amount of research hours spent compared to plasma technology ('CO_2 electrolysis' yields 500 publications on google scholar, 'CO2 plasma' yields 1,500, although this is by no means a comprehensive comparison). Low temperature plasma and atmospheric pressure plasma achieve similar results regardless. Low pressure plasma adds complexity and smaller space-time yields, but achieves higher energy efficiency at similar conversion.

5.3.3   Summary

The mk 1 and mk 2 reactors were successfully used to split CO_2. The mk 2 performed significantly better and reached an energy efficiency of 43% while at the same time reaching 27% conversion. In comparison to similar gliding arc or glow-to-arc transition plasmatrons, a combination of factors can be seen that could explain this performance leap: Coaxial reactors, laminar gas flow, gas quenching, catalysts and axial magnetic fields all have previously lead to high efficiency. The literature review that went into this work did not yield any paper that combined all these approaches. However, the gap between literature results and the results achieved in this study seems to be very large. Thus, reproduction of the observed results is much advised. Gliding arcs appear to be on par with microwave plasmatrons when it comes to technology integration and favourable properties for CO_2 utilization. On one hand, microwave plasmatrons have been scaled to higher power before, on the other hand their drivers reach lower theoretical maximum energy efficiency.

Plasma technology yields competitive results to electrolysis and thermochemical approaches. It will certainly be the technology of choice in some applications. It is scaleable and requires no catalysts or other rare earths, while using a simple overall process. However, the place for plasma technology may not be the pristine splitting of pure CO_2. After all, the fact that no catalysts are used enables a wider set of chemical processes - ones that have the potential to be less energy intensive, albeit less straight forward than a simple stoichiometric reaction of pure gases. Two examples of such processes are treated in the following chapters.

Chapter 6

Energy Efficient CO Production Processes

The direct splitting of CO_2 is an energy intensive procedure. However, the use of plasma has the potential to unlock alternative process routes, that are hard to implement with catalytic or electrolysis systems. Two such processes were briefly studied for their general feasibility in the scope of this work. One will be researched at the ipv-EES (institute for photovoltaics, chair for electrical energy storage systems) as successor to this work in the project *BlueFire*, and will hence be referred to as *BlueFire*-process. Its aim is to use the waste heat of a DC-atmospheric pressure discharge reactor to enable a second, heat utilizing reaction that releases CO_2 from a solid. The burning of lime to cement and CO_2 is one such reaction. The CO_2 is then split in the same reactor inside the plasma.

The second process highlighted in this chapter will be referred to as the *Cyclize*-Process, named after the spin-off that aims to develop the technology in cooperation with the ipv-EES. Key is again the use of waste heat generated in a plasma driven reaction. However in this instance, the reaction enabled by waste heat is the pyrolysis of organic waste materials such as plastic waste. An organic gas is formed, that is mixed with steam or CO_2 and then reformed to syngas in the plasma.

Both of these processes have been claimed as inventions at the University of Stuttgart in the scope of this work and subsequently filed for patenting, together with a reactor design which enables them. This chapter lines out the fundamental design idea and the experiments that have been performed thus far for validation.

6.1   Calcium Carbonate Reduction: The *BlueFire*-Process

The *BlueFire* process makes use of the fact that waste heat in plasma systems is produced at high temperature. Upwards of $900°C$ are necessary to convert limestone - calcium carbonate, $CaCO_3$ - into portland cement - calcium oxide, CaO. The process releases CO_2:

$$CaCO_3 \longrightarrow CaO + CO_2 \,|\, \Delta H_0^r = +178\,kJ\,mol^-1 \tag{6.1}$$

The released CO_2 can in turn be converted to carbon monoxide and oxygen in the plasma. In section 5.2 it was demonstrated that this can occur at up to $\eta_{max} = 43\%$ energy efficiency. Under the assumption that most of the remaining energy is released either as heat or radiation, the heat energy released when splitting CO_2 is at least

$$Q = \frac{1}{1 - \eta_{max}} \Delta H_0^r(CO_2 - splitting) \approx 375\,kJ\,mol^{-1}. \tag{6.2}$$

This is more than enough to heat a stoichiometric amount of calcium carbonate to its decomposition temperature and beyond. This works in principle with every solid or liquid that releases CO_2 when heated. The process thus produces two valuable compounds: Carbon monoxide and calcium oxide. Furthermore, when the burnt cement is used to recapture more CO_2 from the atmosphere, a CCU-process can be realized. A similar approach to CCU has been demonstrated previously, e.g. by Li et al. [131] who used a DBD to release CO_2 from hydrotalcite or by Giammaria et al. who studied the decomposition of Calcium Carbonate in a DBD [132]. Recapturing of atmospheric CO_2 can be implemented by introducing calcium oxide into an alkaline solution of water and generating a large surface area between the solution and the atmosphere. This is the basic principle of operation of a caustic absorber, which are a well established technology. Plasma-based firing of cement in a two-in-one reactor is not. For this reason, preliminary experiments were performed to determine the feasibility of the process.

6.1.1   Experimental Results and Interpretation

Dry and finely powdered calcium carbonate was placed inside the mk 2 plasma reactor instead of the ZrO_2-packed bed. In order to improve heat transfer from the plasma to the solid, a flush stream of nitrogen was used at a flow rate of $\dot{V} = 1\,SLM$. The reactor was run at an applied power of $P = 200\,W$ until it overheats, which happens after 5 minutes with the critical factor being the glass section which is not cooled. A gas sample is taken just before the reactor is turned off and analyzed with the GC. During operation, the amount of CO_2 and CO in the exhaust gas was measured in real time using the NDIR sensors. Multiple runs of the reactor are performed with the same Calcium carbonate sample.

The plasma can be observed to produce orange light emission during operation. This is due to calcium ions that are evaporated at a low rate from the solid. A CO_2-concentration of $c_{CO2} = 1\%$ can be detected in the exhaust gas. In addition, traces of carbon monoxide can be detected. The low conversion of calcium carbonate to CO_2 is expected, due to the small reactor volume that allows for quick heat dissipation. Furthermore, the low conversion to carbon monoxide is also expected, since the gas produced by the discharge is not passed through the discharge volume again, but instead swept out of the reactor by the nitrogen stream. The concentration of CO_2 and CO decreased slightly in subsequent runs using the same sample. Nonetheless, the experiment gives a qualitative proof for the desired process. In order to ensure that calcium oxide has been formed successfully, the treated solids were analyzed. When the reactor was dismantled and the sample extracted, the sample powder seems to be sintered slightly. This is especially true for the upper layer that was directly exposed to the plasma. Under slight mechanical stress the sample returns to its powder form. It can be concluded that the hot gas could not penetrate into deeper layers of the solid after the sintering occurred, which explains the decreasing CO_2-concentration is subsequent runs. A few hundred milligrams of the treated calcium carbonate sample are dissolved in $20\,ml$ of water and the pH is measured using indicator paper. This gives a result of 12-13, while a control sample yields a pH that is close to neutral, thus confirming that calcium oxide was formed that reacts to calcium hydroxide in solution. The overall reaction that took place is

$$CaCO_3 \longrightarrow CaO + CO + \frac{1}{2}O_2 | \Delta H_0^{BlueFire} = 461\,kJ\,mol^{-1}. \qquad (6.3)$$

Under the assumption that specific energy input necessary for CO_2 splitting does not increase as compared to the experiments in chapter 5, the energy efficiency for the whole process increases. Previously, $W_{min}(CO_2 - splitting) = 660\,kJ\,mol^{-1}$

were required to split the CO_2. This could in theory lead to a process efficiency of

$$\eta_{max,theoretical}(BlueFire) = \frac{\Delta H_0^{BlueFire}}{W_{min}(CO_2 - splitting)} = 70\% . \tag{6.4}$$

6.1.2 Proposed Improvements

While qualitative results could be generated, no significant amounts of product were obtained. The generation of quantitative results is not reasonable using the mk 2 reactor, since it is not optimized to transfer heat to the solid packed bed, but rather to dissipate it instead. An optimized reactor design is proposed based on the findings of the experiment, that should meet the following criteria:

- In order to use the product gas, no sweeping gas should be employed. Instead, a gas circulation can be used to transfer heat. A part of the product gas stream is thereby fed back through the plasma reactor as a heat carrier.

- The reactor must be thermally insulated and have a higher nominal power in order to allow for efficient heating of the material.

- A rotary kiln or similar should be implemented to allow for mixing of the solids and prevent sintering.

- A pump must be included to move the working gas through the reactor. It should either use a simple countercurrent heat exchanger or be temperature resistant in order to prevent overheating by the circulating gas.

- An oxygen absorber must be included either in the gas feedback loop or the reactor volume in order to extract the produced oxygen.

The design is illustrated in fig. 6.1. Depending on the process in which the product gas should be used, remaining CO_2 in the product gas can be extracted with an adequate technology and fed back into the reactor. This could be acheived by using a part of the previously produced Calcium Oxide. Extraction might not be necessary if a syngas reaction or a water-gas shift reactor is fueled with the product gas. Products leave at atmospheric pressure and can be analyzed downstream to measure performance. The flow rate of gas should also be measured using a method that can compensate for varying gas composition. An oxygen absorber based on pressure or temperature swing should be considered in the design. It would need to be placed in the feedback gas loop c). The reactor vessel can be manufactured

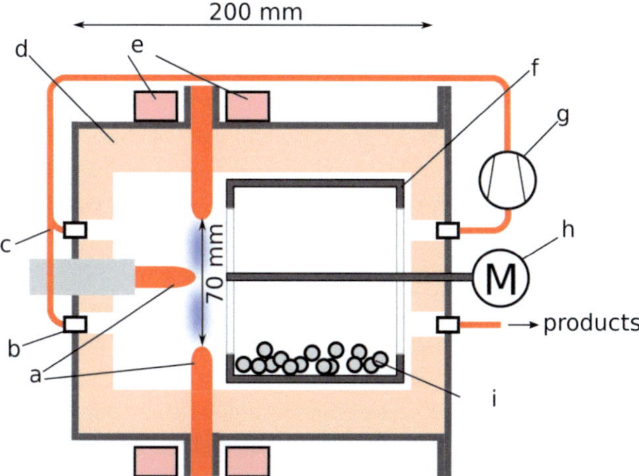

Figure 6.1: Proposed design of an optimized plasma reactor for treatment of solid material in batches. The electrode assembly a) is horizontal. Gas enters through inlets b) after being circulated through pipes c). The reactor is insulated by ceramic foam d) while magnets are places on the outside e). A rotating kiln f) holds the solids i). A pump g) circulates the working gas, a motor h) rotates the kiln.

from 304 or 304H stainless steel (EN 1.4948). This steel is non-magnetic, which is important since the permanent magnets are placed outside of the reaction vessel. Temperature stability is given for up to 900°C, which should suffice for the vessel [133]. For the rotary kiln, 310 stainless steel (EN 1.4835) should be used because of its higher temperature rating (1150°C continuous) [133]. 310 stainless is typically used for furnace parts and also non-magnetic. Slight weakening of the material or creep deformation is not an issue when the reactor is operated at ambient pressure. Piping needs to be made from stainless steel as well. When the reactor walls are insulated on the outside, copper gaskets need to be used for sealing since other compounds are not viable at high temperature. If a temperature of the vessel below $300°C$ can be maintained, silicone or Teflon seals can be used. An insulation made from either aluminum oxide foam or similar oxide or nitrite ceramics is proposed.

6.1.3 Design Principles and Example

The design principles for the proposed reactor are illustrated and accompanied by calculations of an example device in this section. Using the results obtained in section 4.3, the necessary electrode gap for a desired power can be calculated: At a magnetic field of $B = 30\,mT$ a field of $E = 120V\,mm^{-1}$ can be obtained at $I = 200\,mA$ and a strain factor of $\epsilon = 1$. Consequently, a gap of $30\,mm$ should be sufficient to achieve a power

$$P_{Plasma} = I_{discharge}\,E(I, B)\,d\,\epsilon = 720\,W \,, \tag{6.5}$$

while higher current and strain factor can further increase this value beyond the kW-scale. However, no data for the behavior of the discharge at higher current is available yet and average strain depends on the utilized power supply. At this power, a maximum of $1.1\,SLM$ of carbon monoxide is produced when the same energy efficiency of 43% as previously measured could be achieved. The distribution of the applied power between gasification and0 CO_2 splitting is a problem with more than one variable and cannot be determined without extensive measurements. Two general areas of operation are possible for the reactor: If more Calcium carbonate is decomposed into CO_2 than can be split, the CO_2 conversion remains below equilibrium. If less is decomposed, CO_2 conversion should reach equilibrium. The operation point of the proposed reactor can be estimated by assuming an energy efficiency and conversion for CO_2 splitting and then tracking heat flux through the system.

The thermal conductivity of high temperature insulators can reach below $\lambda = 0.04\,W\,K^{-1}m^{-1}$. At an approximate reactor diameter of 120mm and length of $200\,mm$, it has a surface area of $A = 0.06\,m^2$. Using $t = 4\,cm$ of insulation the heat lost through the mantle is

$$Q_{loss,mantle} = \Delta T\,\frac{A\,\lambda}{t} = 0.06\,W\,K^{-1}\,\Delta T. \tag{6.6}$$

At a reactor temperature of $1000\,°C$ this leads to a heat loss of $Q_{loss,mantle} = 60\,W$. Heat losses through metal inlets and the rotary shaft should be expected to at least double this number. Assuming the reactor generates carbon monoxide at an energy efficiency of 43% as achieved in section 5.2, the calculated yield is $1.1\,SLM$ of carbon monoxide. The overall product gas flow at a conversion of 27% is approximately $\dot{V}_{product} = 4.6\,SLM$ and the amount of generated CO_2 has to be at least

$\dot{V}_{CO2,generated} = 4.0\,SLM$. To produce this amount of gas from calcium carbonate, a heat flux of $466\,W$ is necessary. However, after splitting CO_2 and subtracting heat losses the available heat energy is only

$$Q = (1 - \eta)P_{plasma} - Q_{loss} = 410\,W - 180\,W \approx 230\,W\,. \tag{6.7}$$

Thus less CO_2 is produced than in the stoichiometric case. In consequence, the gas remains in the reactor longer and conversion can reach a value closer to thermal equilibrium.

This calculation illustrates, that heat losses in the reactor and piping must be minimized to obtain a good result. Otherwise, too little heat energy will be available to drive the decomposition of calcium carbonate. When the circulation pump cannot be run at the elevated temperature, a heat exchanger must be used. At a recirculation ratio of $R = 2$, the gas flow can reach a maximum of $V_R = 10\,SLM \approx 0.3\,g\,s^{-1}$. At $c_w \approx 1\,kJ\,kg^{-1}\,K^{-1}$ and a temperature difference between reactor outlet and inlet of $\Delta T = 200\,K$, and a medium heat capacity for the gas mixture of $c_p = 1000\,kJ\,kg^{-1}$ the heat loss is

$$Q = \Delta T\,\dot{m}\,c_p \approx 60\,W. \tag{6.8}$$

The results also illustrates the crucial role of an oxygen absorber: Equilibrium conversion will ultimately be reached since fewer CO_2 is emitted than would be converted ideally. With lower oxygen concentration the equilibrium concentration of carbon monoxide increases [108]. The oxygen absorber serves as a laboratory alternative to oxygen scrubbing technologies that are not yet commercially available like selective high temperature oxygen ion conductors.

6.2 Plasma Waste Reforming: The *Cyclize*-Process

Dry plastic waste is a very energy dense material, at up to $35\,MJ\,kg^{-1}$ even more energy dense than crude oil. For this reason it is often used as an energy source in waste power plants when it can't be recycled. Since the chemical industry is turning away from fossil fuels toward electrical chemistry it makes more sense to bring plastic waste to material use instead. This can potentially alleviate the enormous energy requirements for electrical chemistry. The *Cyclize* process is aiming to achieve this goal.

There have been previous attempts to waste pyrolysis using plasma. The goal was either hydrogen generation or cleaner energy production from waste. However most seemed to use a similar approach: A thermal plasma is generated in an inert gas or air. This gas is heated to as high of a temperature as possible and then released into a reaction vessel with the plastic waste [134–137]. The plastic waste is partly converted to energy rich methane, acetylene, or hydrogen, while carbon and other residue stays behind. The inert gas makes material use difficult while thermal plasma consumes a high amount of energy. These two points should mainly be improved: A non-thermal plasma like an atmospheric pressure DC discharge could lead to higher energy efficiency. Igniting the plasma directly in the working gas or rather in the reactor vessel leads to higher heat transfer and does not introduce additional gas. Thus, the goal is to heat a packed bed of waste material using waste heat generated by the plasma. Thereby gaseous hydrocarbons are formed. On the example of polyethylene with the repeat unit [CH$_2$] [138]:

$$[CH_2](s) \longrightarrow CH_4,\ C_2H_2,\ C_2H_4,\ C_2H_6,\ C,\ H_2 = CH_2(g)\,|\,\Delta H^R > 0\,. \tag{6.9}$$

The organic gases are then circulated back into the plasma and reformed with an oxygen containing gas like steam or CO_2. Reforming of hydrocarbons with CO_2 is called dry reforming. On the example of gases produced as an intermediate product from CH_2, the average reaction is:

$$CH_2(g) + 2\,CO_2 \longrightarrow 3\,CO + H_2O(g)\,|\,\Delta H^R > 0. \tag{6.10}$$

The overall reaction is thus

$$CH_2(s) + 2\,CO_2 \longrightarrow 3\,CO + H_2O(g)\,|\,\Delta H^R = 269\,kJ\,mol^{-1}, \tag{6.11}$$

using an enthalpy of formation for each CH$_2$ unit of $H_f^0(CH_2) = -55\,kJ\,mol^{-1}$, based on the repeat unit of polyethlyene. Two preliminary experiments validate, if this reforming reaction is possible in a non-thermal gliding arc reactor. Another approach is the steam reforming of hydrocarbons. Here, water is used as oxygen containing gas to produce mostly hydrogen:

$$CH_2(s) + 2\,H_2O \longrightarrow 3\,H_2 + CO_2\,|\,\Delta H^R = 146\,kJ\,mol^{-1}, \tag{6.12}$$

6.2.1   Experimental Results and Interpretation

First, the reforming of hydrocarbons was tested. A flow of $\dot{V} = 1.4\,SLM$ is fed through the mk 2 plasma reactor with packed bed at an applied power of $P = 140\,W$. The gas consists of equal parts of CO_2 and methane. The product gas was only analyzed with the GC, since residual carbon in the gas could destroy the NDIR gas sensors. A sample is taken after 2 minutes of operation.

During operation it is apparent, that the plasma is less stable than during operation in pure CO_2. Fluctuations in the swept volume occur and the discharge channel is often pinned in place, The colour of emitted light is orange instead of the usual blue, which is typical for soot in a carbon rich environment. The reason for both phenomena is revealed after the plasma reactor is turned off: Residual carbon from the reaction can be found on virtually all surfaces of the reactor. Since carbon is conductive, this can alter the path of the discharge. Soot can also inhibit reignition of the plasma when a voltage multiplier is used, because it prevents the buildup of high voltage due to leak current flowing between the electrodes through the carbon layer. The reactor can be regenerated by running it with pure CO_2, which slowly reacts with the carbon residue on all hot surfaces due to the Boudouard reaction. The abundance of residual carbon also manifests in the composition of the product gas. The composition is shown in fig. 6.2.

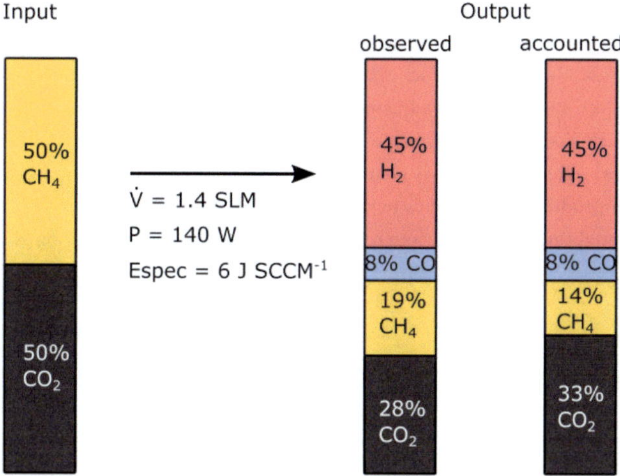

Figure 6.2: Gas concentration on inlet and exhaust of the mk 2 plasma reactor in the performed dry reforming experiment, alongside accounted product distribution according to reactions (6.14) and (6.14). Process parameters are shown in the centre.

The achieved products can be explained by two parallel reactions that take place.

The first is the desired dry reforming reaction:

$$1)\ CH_4 + CO_2 \xrightarrow{X_1} 2CO + 2H_2 \,. \tag{6.13}$$

The second is the pyrolysis of methane:

$$2)\ CH_4 \xrightarrow{X_2} C + 2H_2 \,. \tag{6.14}$$

Other reactions may occur, although no significant concentrations of other products were found in the GC analysis. All products which are gases below $150\,°C$ and have lower interaction with the column than CO_2 should in theory be detectable in the utilized GC as extra peaks. The conversion of both reactions, X_1 and X_2 can be calculated from the concentrations in the product gas $c(H_2)$ and $c(CO)$ without knowing the flow rate downstream of the reactor to be $X_1 = 0.1$ and $X_2 = 0.5$. Details for this calculation can be found in appendix B. The calculated results show a slight discrepancy with the measured concentrations of CO_2 and CH_4. One explanation is that the formation of water and residue carbon occurs, which can not be measured. Gaseous C2 compounds and above are not expected, since no additional peaks in the GC spectrum were identified. The concentration of CO_2 and CH_4 in the feed flow was only adjusted by using an analogue mass flow meter. Thus, variations in feed gas concentration could also explain the difference between measurement and calculation. The results can be compared to the equilibrium concentration of product gases that is expected at different temperatures. Fig. 6.3 shows the simulated thermal equilibrium concentrations as described in section 2.2.4.

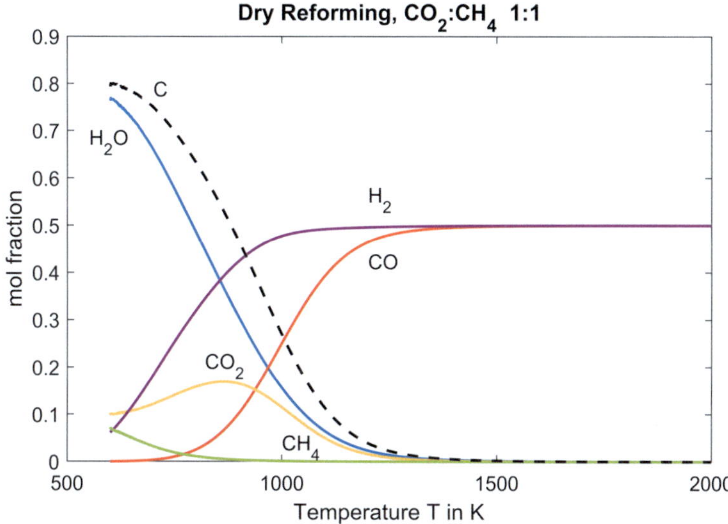

Figure 6.3: Simulated equilibrium products of methane dry reforming using a 1:1 ratio. Note that carbon as a solid is not included in the mol fractions.

It can be seen that a high hydrogen to carbon monoxide ratio is achieved at low temperatures, where methane pyrolysis is the dominant process. A hydrogen concentration of $c_{H2} = 45\%$ after water condensation is achieved at $660\,K$. However, almost no carbon monoxide is formed in this temperature regime. Under the assumption that thermal equilibrium is reached, a possible explanation for the measured conversion is that two different temperature regions dominate the process: Within the discharge channel, carbon monoxide is formed at high temperatures. In the colder afterglow and on the catalyst bed, methane is pyrolysed. Both product streams mix to produce the observed product concentrations.

The results also suggest, that methane pyrolysis contributes significantly more to the overall reaction than dry reforming. This is a problem for syngas production. It can be negated by increasing the concentration of oxygen rich gases like CO_2 or water vapour in the supplied gas. Furthermore, the formation of carbon can be prevented when no cold surfaces are present - at elevated temperature, carbon residue reacts with reducing gases more easily. Thus, the avoidance of cold spots and generation of sufficient temperature is an important design criterion for an optimized reactor. Under the assumption that residue carbon is present in excess, the yield of carbon monoxide is dictated by temperature due to the Boudouard equilibrium.

The second performed experiment is the gasification and reforming of organic solids to carbon monoxide and water. Polystyrene was chosen for the test due to its abundance in the laboratory and the fact that it is one of the most employed polymers worldwide. At elevated temperature, polystyrene can depolymerize to styrene, which is then present as a vapour that can condense below $145°C$. A sample of a few grams is placed in the mk 2 plasma reactor instead of the packed ZrO_2 bed. A flow of 1.25 SLM of CO_2 is employed, the applied power is $P_{plasma} = 165\,W$. Just as in the previous experiments, the working gas is not recirculated. After two minutes of operation, a gas sample is taken and analysed with the GC. The results are confirmed by in-line NDIR sensors, which might not have been the best idea in retrospect. The discharge color once again is mostly orange, although blue light emission is still observed inconsistently. The discharge is also less stable in geometry than when pure CO_2 is used. The product gas consists of approximately $60\%\,CO_2$, $40\%\,CO$ and less than 1% hydrogen. The gas concentration changes as the solid sample is gasified over the coarse of several runs with a duration of 2 minutes each. Just as in the dry reforming test, residual carbon is abundant. The residual solid is dry and loosely packed. The precipitation of soot is mostly limited to the volume around and below the electrodes, which creates no risk of an electrical short circuit. The lack of hydrogen in the product suggests, that the water gas shift reaction towards water is nearly completed, which is typical for high temperatures. The amount of water that is formed in the reactor is too small to be measured. However, condensation of water and possibly organic liquids can be observed in the tubing and metal parts downstream of the reactor.

The utilization of remaining carbon can be realized by increasing the reactor temperature and gas recirculation, which facilitates the Boudouard reaction. The fact that no organic gases or vast amounts of organic liquids are among the products is somewhat surprising, considering the heat of the reactor could drive the decomposition of the polymer. There was also virtually no oxygen present in the product gas. Also, the obtained yield of CO is lower than could be expected by the ratio between power input and reaction enthalpy, especially when compared to the yield achieved previously in the splitting off pure CO_2. This suggests, that a major reaction path is the splitting of CO_2, while hot oxygen subsequently reacts with the polystyrene.

It must be concluded, that the mk 2 reactor is not well suited to the application and that gas circulation is a key feature of an improved reactor.

6.2.2 Proposed Improvements

Compared to the *BlueFire*-process, lower temperatures are theoretically necessary to facilitate the gasification of organic matter. Lower temperatures might even lead to higher overall conversion to organic gases instead of carbon and hydrogen. However, it is expected that solid carbon will be left after gasification under any circumstances. This can be seen as an advantage: Pyrolysis-made graphite is a valuable resource. Extracting carbon furthermore reduces the CO_2-footprint of the technology. However, the carbon will be mixed with non-volatile components such as metals, which makes both material use or disposal difficult. The carbon could instead be utilized by increasing the temperature in the reactor before extraction of the ash. The Boudouard-equilibrium and water gas shift will work together to gasify the carbon at high temperatures.

A reactor that moves gas and solids in countercurrent would be well suited for this application. The solids should move towards the plasma reactor, so that they are subject to the highest temperatures before being discarded. The extracted gas should be quenched before being extracted to prevent carbon deposition. Furthermore, no cold spots should be found in the piping of the reactor and the reactor itself. Since steam can be used as input gas and appears as a product, condensation would occur there. On the contrary, it can be beneficial to extract water or CO_2 from the reactor to shift the water-gas-shift equilibrium, as will be discussed in the next section. The extraction of water can be implemented by a condenser in the circulation line. Extracting CO_2 is more challenging, since an absorber is at risk of destruction through organic compounds and halogens in the gas.

6.2.3 Design Principles

The calculations concerning heat availability established previously for the burning of cement apply also in the reforming of solid hydrocarbons. However, an additional concern is the selectivity, since more than one product can be formed. The desired product of the *Cyclize* process is either hydrogen, carbon monoxide or a fixed ratio of both to be used as Syngas. To produce these products, either carbon dioxide or steam is used as a feedstock. In the preliminary experiments it was found that multiple processes, namely dry reforming and pyrolysis contribute to the overall conversion.

Products are subject to thermal equilibrium. Three reactions influence thermal equilibrium in a reactor where steam and carbon monoxide are present: The before mentioned Boudouard equilibrium, the water-gas-shift reaction and the reforming of methane. Methane was included into this study, since it is the most stable hydrocarbon. Its presence can therefor be used as an indicator for the presence of hydrocarbons in general [139]. Methane is also a major product of thermal decomposition of polyethylene [138].

In this reaction system, temperature is the key parameter when a certain product is desired. While concentrations of gases inside a reactor might not reach thermal equilibrium before the product is extracted, the equilibrium gives an upper bound on product concentration. Another unknown in this respect is the plasma. A reforming reaction taking place inside the plasma might reach conversions exceeding chemical equilibrium. However, for the sake of simplicity, the thermal equilibrium is analyzed first to derive design criteria for a waste reforming reactor.

Carbon Monoxide by Dry Reforming

Fig. 6.4 shows the equilibrium concentration achieved in dry reforming of polyethylene at varying temperature.

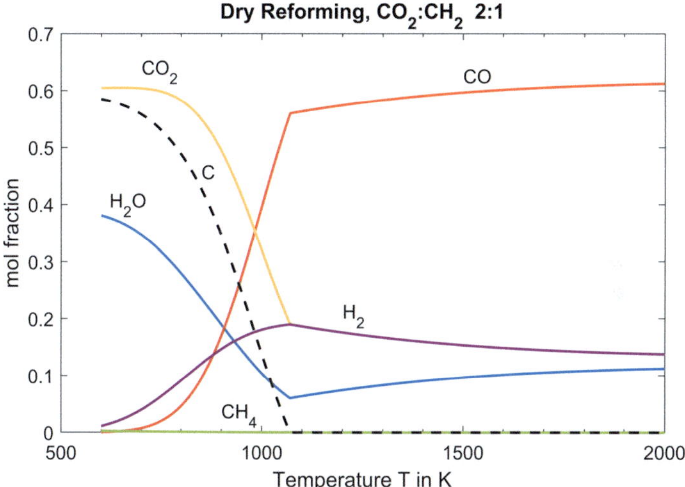

Figure 6.4: Simulated thermal equilibrium concentration of gases produced in the 2:1 dry reforming of polyethylene. Note, that carbon as a solid is not considered in the mol fractions.

As expected, at a temperature below $1100\,K$, carbon is formed in large amounts. Above this temperature, carbon monoxide is produced with a concentration of roughly $c_{CO} = 60\%$, almost independent of temperature. However, the ratio of carbon monoxide and hydrogen is still varying above this temperature, which is illustrated in fig. 6.7. Concentration curves show a sharp edge at a temperature of $1100\,K$. This corresponds to the temperature at which no solid carbon is left. The reason for this sharp edge is, that carbon is not included in the employed equation for the Boudouard equilibrium. In reality, low amounts of carbon will persist beyond this temperature. No methane remains in the product gas after dry reforming at any temperature when thermal equilibrium is reached.

If carbon monoxide is desired as a product, the WGS equilibrium can be influenced to favour it. This can be achieved by using an excess of carbon dioxide, the simulated results of which are shown in fig. 6.5.

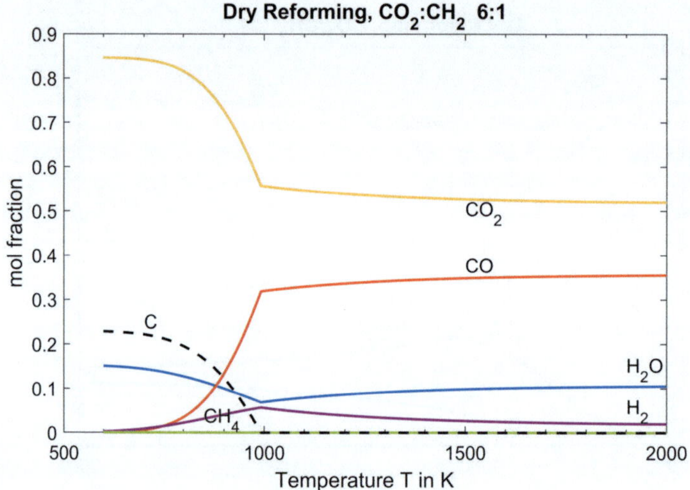

Figure 6.5: Simulated thermal equilibrium concentration of gases produced in the dry reforming of polyethylene at an excess of carbon dioxide. A 200% excess of CO_2 is used. Note, that carbon as a solid is not considered in the mol fractions.

The concentration of CO lowers, because it is diluted by excess CO_2. Nonetheless, the CO-to-H_2 ratio is higher. This approach is questionable, because CO_2 is difficult to separate from the product gas. A pressure swing absorber would need to be dimensioned excessively large to obtain reasonable purity.

Removing water leads to even higher concentration of carbon monoxide in the product gas, which is shown in fig. 6.6. Up to 80% CO are obtained, while the concentration of hydrogen and CO_2 both are in the range of 10%. Combining both approaches would lead to even higher equilibrium conversion to CO.

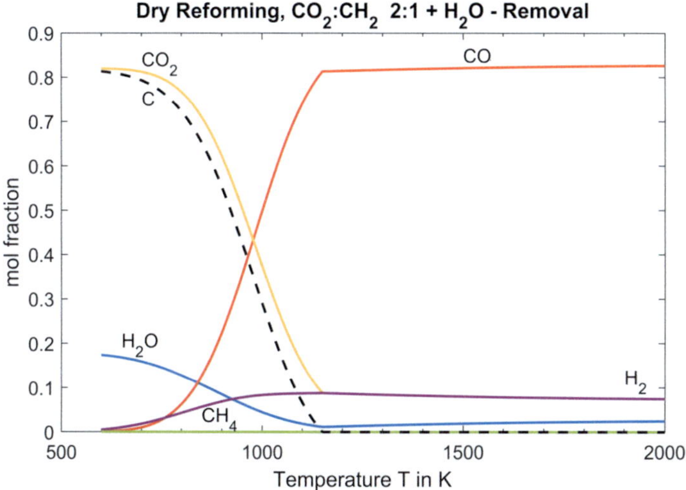

Figure 6.6: Simulated thermal equilibrium concentration of gases produced in the dry reforming of polyethylene when 67% of water are removed. Note, that carbon as a solid is not considered in the mol fractions.

The approaches and their carbon monoxide yield are compared in fig. 6.7. The highest carbon monoxide to hydrogen ratio can be obtained at high temperatures and by using excess CO_2. This approach also results in the highest carbon monoxide yield of close to 95% based on the stoichiometric conversion of polyethylene. Water condensation instead of excess CO_2 leads to similar results, but is easier to implement. Pure carbon monoxide is used commercially either in methanol carbonylation to produce acetic acid, or in phosgene production. While the first process might tolerate the remaining CO_2 and hydrogen, the second certainly doesn't. Thus, product purification is necessary for phosgene production, but not in other syngas applications. The purification could be implemented by using pressure swing absorption to capture the remaining CO_2 and using a condenser in the recirculation loop of the reactor to extract water vapour.

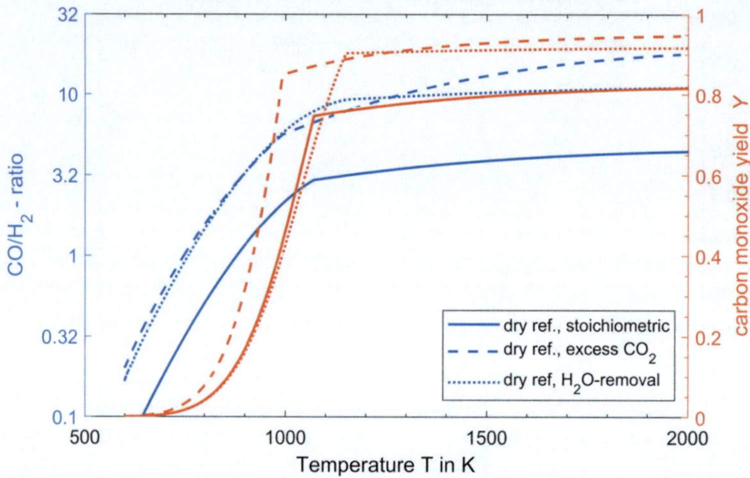

Figure 6.7: On the left axis, the simulated ratio of carbon monoxide and hydrogen produced in the dry reforming of polyethylene is shown, when performed stoichiometrically, with a 200% excess of CO_2, and when two thirds of the steam are extracted from the recirculation stream. The right axis shows the carbon monoxide yield based on stoichiometric conversion.

Hydrogen by Steam Reforming

Pure hydrogen is used as a reagent, but is also gaining attention as a fuel. Since it is used in fuel cells, no carbon monoxide should be present. Water gas shift reactors, methanaization or pressure swing absorption are typically used to increase the purity of hydrogen beyond 99.9%. However, the feedstock gas should be reasonably hydrogen-rich, i.e. above 50%, from the start to decrease the cost of purification. The simulated product composition of stoichiometric steam reforming of polyethylene is illustrated in fig. 6.8. Steam is used as working gas in the plasma.

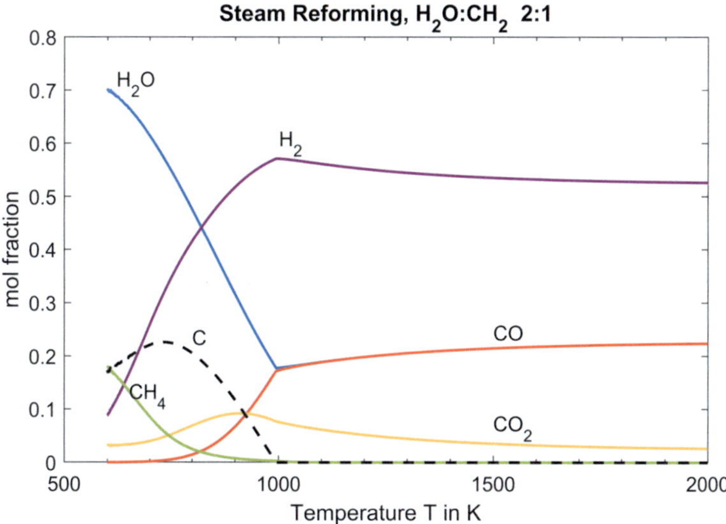

Figure 6.8: Simulated thermal equilibrium concentration of gases produced in the stoichimetric steam reforming of polyethylene. Note, that carbon as a solid is not considered in the mol fractions.

To increase the amount of hydrogen in the product gas, an excess of steam can be used. This is also usually done in water-gas-shift reactors. The excess steam can easily be condensed out of the product gas, which is a contrast to excess CO_2. The product concentrations at equilibrium are illustrated in fig. 6.9. The simulation suggests, that no significant carbon deposition is expected in thermal equilibrium, except in a small window between $700\,K$ and $800\,K$. At temperatures above $1200\,K$, carbon is mostly present as monoxide. Hydrocarbons are fully converted at temperatures above $900\,K$.

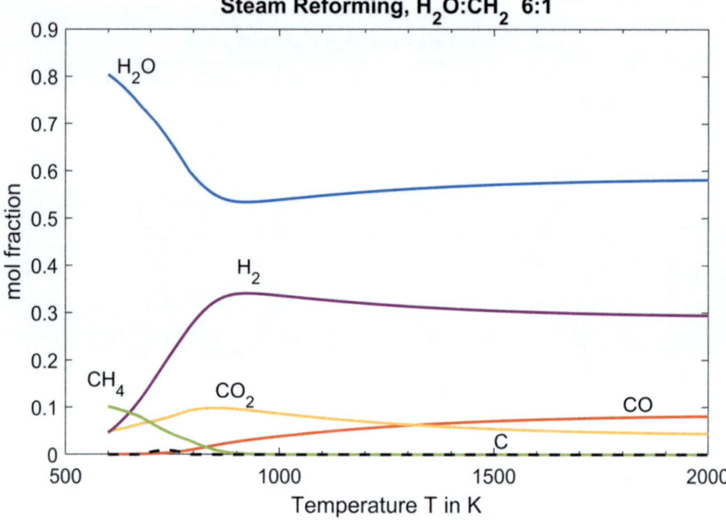

Figure 6.9: Simulated thermal equilibrium concentration of gases produced in the water-rich steam reforming of polyethylene at varying temperature. A 200% excess of steam is used. Note, that carbon as a solid is not considered in the mol fractions.

Alternatively, CO_2 can be extracted from the recirculation gas stream. This could be implemented as pressure swing absorber, although technical challenges are likely. Extracting two thirds of the forming CO_2 results in the equilibrium product concentrations shown in fig. 6.10. The simulation results suggest that CO_2-removal leads to the highest hydrogen yield. Furthermore, the temperature above which carbon deposition is not problematic is also decreased to $1000\,K$. Extraction of CO_2 also decreases carbon dioxide emissions that could occur in downstream processes, e.g. when leftover gas is burned after a Fischer-Tropsch reactor.

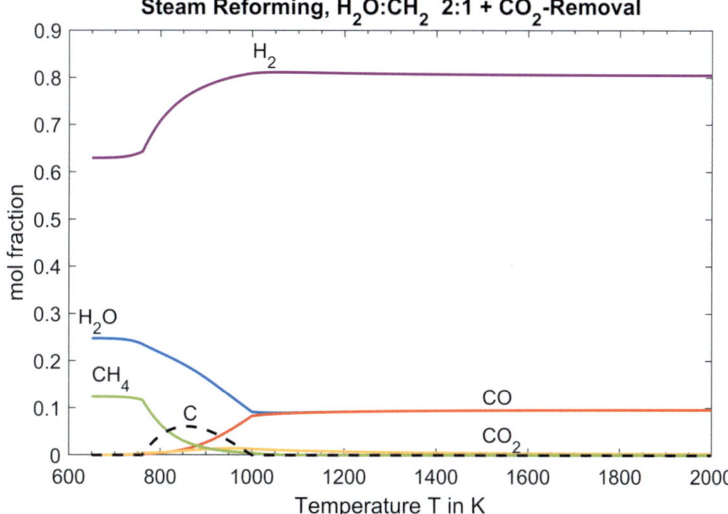

Figure 6.10: Simulated thermal equilibrium concentration of gases produced in the steam reforming of polyethylene while extracting CO_2 from the recirculation stream at varying temperature. Note, that carbon as a solid is not considered in the mol fractions. Two thirds of CO_2 are extracted.

Each approach to steam reforming creates a different ratio of carbon monoxide to hydrogen. They are all illustrated in fig. 6.11. The most consistent and highest yield can be achieved by removing CO_2 from the recirculation gas stream. Operation at moderate temperature with excess water achieves similar results and is expected to be technologically easier to implement. At stoichimetric operation, a hydrogen-to-carbon monoxide ratio of 3.3 can be achieved at significant yield. The best yield of 90% is achieved by steam reforming with excess water at a temperature of $900\,K$. Removing CO_2 from the reactor results in a similar yield at higher temperature.

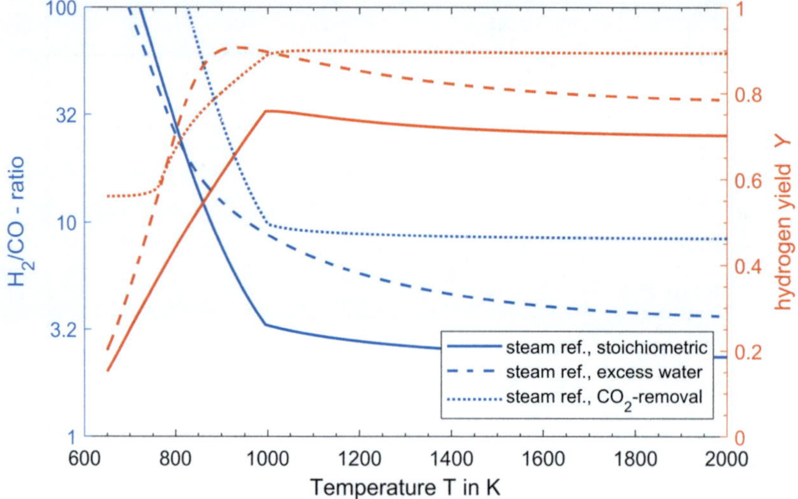

Figure 6.11: On the left axis, the simulated ratio of carbon monoxide and hydrogen produced in the steam reforming process of polyetyhlene is shown, when performed stoichiometrically, with a 100% excess of H_2, and when two thirds of the CO_2 are extracted from the recirculation stream. The right axis shows thy hydrogen yield based on stoichiometric conversion.

In all previous simulations, the total mol number of oxygen exceeds the mol number of carbon. Thus byproducts in the form of either water or CO_2 are inevitable. When exactly the same mol number of oxygen and carbon is used, virtually no byproducts are expected at high temperatures. This can be achieved by using an arbitrary linear combination of two reaction equations:

$$CH_2 + H_2O \longrightarrow CO + 2H_2 \mid \Delta H^R = 187\,kJ\,mol^{-1} \quad (I)$$
$$CH_2 + CO_2 \longrightarrow 2\,CO + H_2 \mid \Delta H^R = 228\,kJ\,mol^{-1} \quad (II)$$

$$(6.15)$$

In effect, syngas with a concentration ranging from 2:1 to 1:2 can be produced. Hydrogen-rich syngas is certainly more interesting commercially, 2:1 syngas can be used for the synthesis of long-chain alkanes in the Fischer-Tropsch process or the synthesis of methanol, making it the worldwide most used composition. The simulated product concentration obtained using equation 6.15 (I) is illustrated in fig. 6.12. It can be seen that the 2:1 ratio is precisely achieved at any temperature above $1200\,K$. Carbon is formed at temperatures of up to $1500\,K$.

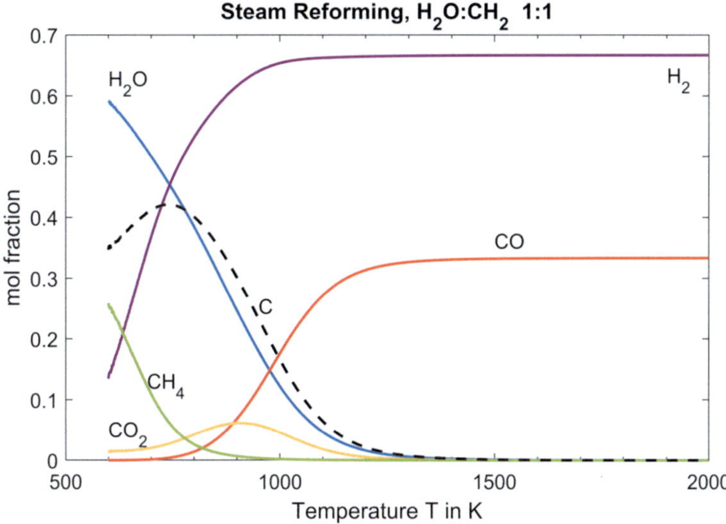

Figure 6.12: Simulated thermal equilibrium concentration of gases produced in the steam reforming of polyethylene according to equation 6.15. Note, that carbon as a solid is not considered in the mol fractions.

The results obtained in simulations of dry- and steam reforming can be used to give design recommendations for a reactor depending on the desired product. All simulations have in common, that the Boudouard reaction dominates the product distribution up to a temperature of $1000 \ldots 1500\,K$. Only the WGS acts above this temperature and its effects are rather limited in comparison. This leads to the conclusion, that the operation temperature of a reactor should be in the range of $1200\,K$ when excess water is present or above $1500\,K$ when it is not. During operation of a plasma waste reforming system, gas that is heated in the plasma and flows over the packed bed is cooled, while the solids are gasified. Thus, depending on geometry, not all parts of the reactor reach the same temperature. The solids should be heated to a temperature that allows full utilization of carbon before extraction, while lower temperatures can be tolerated further from the plasma.

A 1-dimensional model could lead to deeper insights into the design of such a reactor, where different temperatures and concentrations are allowed along the length of a virtual rotary kiln reactor. Implementing such a model exceeds the scope of this work. Including more hydrocarbon gas species along with their chemical equilibria could be necessary to describe low temperatures accurately.

The temperature spread could even be used on purpose in a two step reaction: First, at low temperature hydrogen and carbon could be produced. In a second, high temperature reaction step the remaining carbon is reacted in a dry reforming reaction to produce carbon monoxide.

Additional design considerations can be made in conclusion to the simulation results: The flow of input gases, CO_2 and steam, should be controllable in real time in order to achieve the desired product distribution. A boiler is necessary to produce the required steam. The production of 2:1 syngas according to equation 6.15 (I) uses $187\,kJ\,mol^{-1}$ of energy. This corresponds to a specific energy requirement of $E = 3.7\,kWh\,kg^{-1}$ of polyethylene. At the same time, $1.3\,kg$ of steam is used. The vaporization of this mass of water requires an additional $0.63\,kWh$ of energy, not including the heating of the water up to its boiling point. In an industrial plant this energy could be supplied as waste heat, but in a demonstration plant an electrical boiler should be used. Alternatively, the water could be sprayed directly into the plasma reactor.

Per mol of polyethylene, 3 mol of gases are produced. This corresponds to a gas volume of $1.6\,Nm^3\,kg^{-1}$ of PE. Heat loss limits the amount of solids that can be processed in a reactor at any given applied power. The heat loss calculation can be carried out in a similar fashion as in the previous section for the *BlueFire* process. Since the process temperatures are on a similar order of magnitude, it is expected that about half of the heat is lost on a 1-kW scale. The distribution between energy used for the gasification and the reforming cannot be estimated yet, since the products obtained by gasification are not known.

When pure CO or H_2 are the desired product, a condenser or carbon dioxide scrubber needs to be implemented in the recirculation gas stream according to the performed illustrations. Product gas purification is similar to other syngas processes and should thus not be the scope of future small scale studies. Per kilogram of polyethylene, $1.3\,kg$ of water are formed when full conversion towards carbon monoxide is achieved. A condenser should be dimensioned accordingly. In the steam reforming reaction, $3.14\,kg$ of carbon dioxide are formed at full conversion towards hydrogen, which corresponds to $1.6\,Nm^3$.

The results obtained underline the importance of temperature control and oxygen balance, which help to obtain the desired products and mitigate the problem of carbon deposition on surfaces. This can be a problem especially on surfaces which need to be cooled, such as the electrodes of the plasmatron. Long term studies are necessary to judge the stability of a kW-scale system, which needs to be the next step towards industrial application.

6.3 Summary

The burning of metal carbonates to oxides and carbon monoxide, as well as the reforming of organic waste to syngas could be valueable processes in a carbon neutral future. They both offer the potential for huge reductions in electricity consumption compared to the plasma-based splitting of CO_2. However, both could only be demonstrated qualitatively. A new system, at least on 1-kW-scale is necessary to overcome heat loss and enable the construction of an adequate feed system for solid material. Preliminary experiments gave hints to design goals and potential challenges. Based on these experiments, simulations and prior calculations were carried out. A design proposal was made according to the results. A big challenge for small scale systems is the heat loss, which is mitigated in multi-kW or MW systems. The underlying chemical processes for the reforming of polymers were simulated. Real time control of the feed gas composition in response to product gas composition is an important tool to reach a desired product gas. The ability to selectively remove water or carbon dioxide from the reactor recirculation can help to achieve high yield of the desired product gases. When the oxygen balance is kept neutral, e.g. in a 1:1 steam reforming reaction, carbon deposition can occur at temperatures of up to $1500\,K$. Excess water or carbon dioxide can mitigate this problem.

Chapter 7

Feasibility Calculations

Three major processes have so far been presented in this thesis: The splitting of CO_2, burning of carbonates to oxides and carbon monoxide, and the reforming of plastics. All three processes offer the possibility to reduce carbon emission in the production of chemicals and move away from fossil feedstock. However, the industrial implementation of a technology is not usually driven by environmental concerns. Chemicals are made by companies, which need to turn a profit within the constraints given by the free market and politics. Fortunately, the public is well aware of the ramifications of climate change and consequently political measures are undertaken to make carbon-neutral technologies more economic. An example of this is the carbon dioxide certificate system enacted in the EU, in which prices for emissions have risen to a record high, approaching $100 \, €\, t^{-1}$ as of February 2022 [98]. The move away from coal, arguably the most emission-intensive feedstock, together with the precarious political situation in eastern Europe has propelled prices for natural gas to an all-time high of more than 110 € per MWh, which corresponds to approximately 1350 € per ton [97, 140]. This is probably a short time phenomenon, however market developments will influence the choice of technology in the near future.

In this chapter the influence of the price of commodities and energy on syngas production cost is analysed for two plasma processes: CO_2 splitting coupled with water electrolysis and plasma waste reforming. They are compared using literature data to steam reforming of methane, which is the sate-of-the-art thermochemical process, as well as to co-electrolysis, on the example of syngas production for methanol. Methanol is one of the major uses for syngas and hence for carbon monoxide and requires roughly two parts of hydrogen and one part of carbon monoxide by volume.

For the plasma processes meanwhile a simplified plant design is proposed that can be analyzed as well. The mass flow of feedstock and chemicals though the plants is calculated for varying performance parameters like conversion and energy efficiency. Different scenarios are given for performance; a 'pessimistic' scenario assumes no improvement over demonstrated performance while different scenarios for performance improvements are given. For each scenario, the dimensions of plant components and associated capital expenditures are calculated. Finally, costs for feedstock, energy, personal, CO_2 emissions, maintenance and capital are added to obtain the levelized costs of production. This approach also allows the judgement of direct and indirect CO_2 emissions. The goal is to gain an understanding of the market conditions that are necessary for plasma technology to be implemented successfully on an industrial scale. An extended goal is to give hints how CO_2 emissions and added product cost could be balanced. All results were first generated to judge the viability of waste reforming in the scope of a start-up Cyclize and later refined for this dissertation, but have not previously been published.

7.1 CO_2 splitting and Water Electrolysis

This section discusses the economic viability of syngas production by water electrolysis combined with plasma based CO_2 splitting. The proposed process is illustrated in fig. 7.1.

Plant costs comprise water electrolysis, a plasma unit, oxygen absorber and CO_2 scrubber. According to Nguyen et al. [141], water electrolysis can be used to produce hydrogen at a cost of 3000 € per ton on a scale similar to the here examined 100 tons of syngas per day. They conclude that 36% of the costs are linked to electricity, with the remainder accounted for mainly by capital expenditures. Water electrolysis is considered carbon neutral in this study and hence levelized cost of hydrogen only depend on electricity cost. Plant costs were adapted and scaled from Nguyen et al., who evaluate a 100 MW installation to cost approximately 145 million €. This results in 62 million € of capital expenditures for the electrolysis plant alone according to the scenario marked as 'target' in table 7.1. The plant efficiency for an alkaline electrolysis plant is taken to be $\eta_{electrolysis} = 62\%$. In a master thesis performed at the institute [142] it was found, that a high temperature electrolysis could utilize plasma waste heat to reach higher efficiency. This approach was not

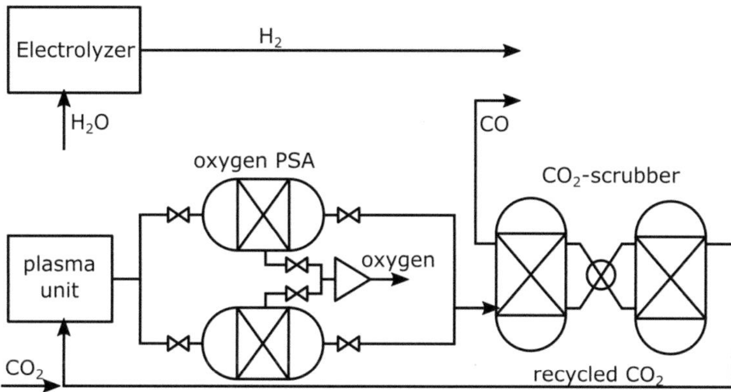

Figure 7.1: Proposed plant configuration for a syngas process using plasma CO$_2$ splitting.

considered here for lack of accurate data on plant costs.

Case	$X_{splitting}$	$\eta_{splitting}$	$\eta_{electrolysis}$	CapEx estimate	installed power
pessimistic	0.3	0.45	0.62	75 m €	61 MW
improved conversion	0.45	0.45	0.62	72 m €	61 MW
improved efficiency	0.3	0.6	0.62	74 m €	55 MW
improved Electrolysis	0.3	0.45	0.75	67 m €	56 MW
target	0.45	0.6	0.62	70 m €	54 MW
optimistic	0.5	0.7	0.9	55 m €	40 MW

Table 7.1: Parameters used for the production cost evaluation of syngas production by plasma based CO$_2$ splitting combined with water electrolysis shown in fig. 7.1.

The capital costs for the plasma process are estimated according to a $3\,kW$ gliding arc plasmatron that is under construction at the institute of photovoltaics as part of this work at the time of this writing. The efficiency $\eta_{splitting}$ and CO$_2$ conversion of the plasma reactor $X_{splitting}$ in the 'pessimistic' scenario are similar to the values measured in chapter 5. In addition, separation of the product gas is necessary. Oxygen absorption can be performed using pressure swing absorption (PSA). Other technologies might lead to better results, such as high temperature oxygen conductive membranes, but no information on these technologies on industrial scale could be found. PSA is assumed to require an electrical energy of 300 kWh per ton of oxygen [143]. The energy costs are assumed to be constant, regardless of the starting concentration of oxygen in the working gas. The oxygen capture ratio is assumed to be 95%, while the remaining oxygen reacts with carbon monoxide back to CO_2. In effect, this reduces the conversion efficiency of the plasma process by 5% relatively. The amount of oxygen that needs to be extracted is constant across scenarios at 50 tons per day (TPD), which is given by the stoichiometric production of 87.5 TPD of carbon monoxide. The capital expenditures of the oxygen absorber are estimated to be 5.5 million € in the scenario marked as 'target' in table 7.1.

Since conversion is low in plasma based CO_2 splitting, a CO_2 scrubber is included in the system, which allows the recycling of the gas and increases product purity. This reduces both the CO_2 emission and the required feedstock. An amine scrubber is chosen since it is a well established technology. Membrane separators or high temperature carbonate ion conductors might be a more desirable technology, but no data could be found on commercial systems of significant scale. A CO_2 recycling rate of 90% is assumed. The scale of the amine scrubber is calculated in reference to the mass flow of CO_2, $M_{CO2,recycled}$, which in turn is given by the conversion of the plasma system:

$$M_{CO2,recycled} = 0.9\, M_{CO2,emitted} = 0.9\,(1 - X)\, M_{CO2,required}$$
$$= 0.9\,(1 - X)\,\frac{28\,g\,mol^{-1}}{44\,g\,mol^{-1}}\, M_{CO,produced}. \tag{7.1}$$

The feedstock of CO_2 that is used is the difference between required mass and recycled mass of CO_2. The recycling of CO_2 is assumed to cost $0.1\,MWh$ electrical energy per ton in addition to heat energy [144]. Plenty of waste heat is available from the plasma process, thus heat is not considered in the energy balance. Energy costs and capital expenditures for the CO_2 scrubber are adapted from Husebye et al. [144]. The capital expenditures for the amine scrubber used in the scenario marked as 'target' in table 7.1 are calculated to be 5.1 million €. Fig. 7.2 shows the calculated production cost composition of syngas using the described process. It is apparent, that the majority of costs is accounted for by electricity consumption of the plant.

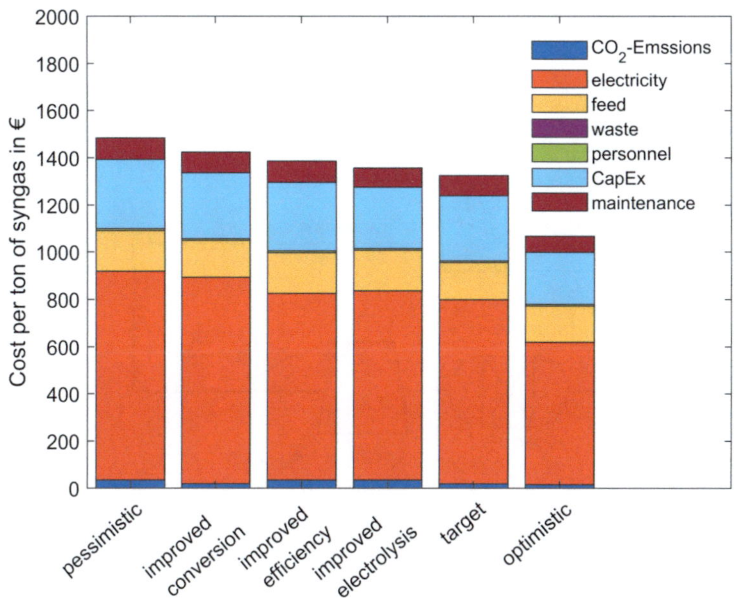

Figure 7.2: Calculated production cost composition of syngas by plasma based CO$_2$ splitting combined with water electrolysis at varying plant performance using the process shown in fig. 7.1.

In the 'target' scenario, 61% of the electricity is used in electrolysis, despite the fact that only one process step is involved as opposed to the three steps necessary for *CO*-production. This ratio does not fluctuate significantly in all scenarios. It is 54% in the 'pessimistic' scenario and 56% in the 'optimistic' scenario - even though the 'optimistic' scenario uses the thermodynamic optimum energy efficiency that can be achieve for alkaline electrolysis. Hydrogen accounts for two thirds of syngas by volume, so it is not surprising that its production uses roughly two thirds of the energy, given that the reaction enthalpy for water splitting and CO$_2$ splitting is similar. Feedstock costs are almost exclusively due to CO$_2$ usage. They vary only slightly throughout scenarios, since the recycling rate of CO$_2$ that is assumed is very high. For the same reason, CO$_2$ emission costs are low.

The high CapEx and resulting maintenance, especially of the electrolyzer, also ac-

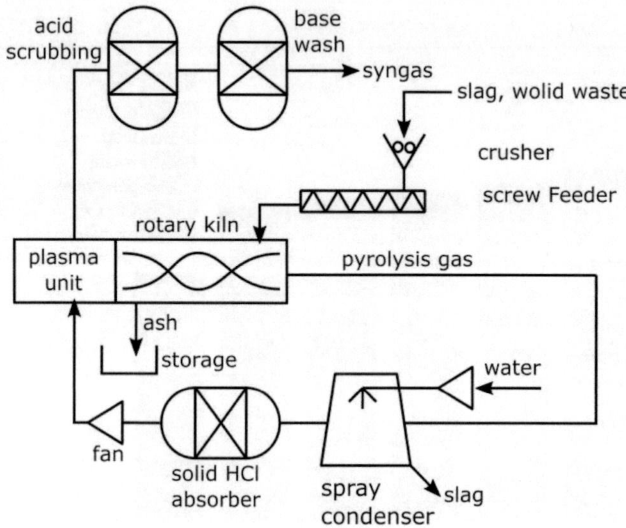

Figure 7.3: Proposed plant configuration for a plasma waste reforming plant according to chapter 6.

count for a considerable percentage of product cost. All remaining costs are insignificant. In a 'pessimistic' scenario, where performance does not increase compared to current results, syngas can be made at a cost of $1480\,€t^{-1}$, while in the 'target' scenario costs are only slightly reduced to $1330\,€t^{-1}$. Generally, improving energy efficiency η_{plasma} has a higher impact than improving conversion X_{plasma}. As discussed before, improving energy efficiency of the electrolyzer $\eta_{electrolysis}$ has the greatest impact. Under 'optimistic' conditions, where both water electrolysis and the plasma process reach close to optimum efficiency, syngas production costs remain at $1035\,€\,t^{-1}$. The 'target' scenario is used in the following comparison to alternative technologies.

7.2   Plasma Waste Reforming

This section discusses the economic viability of a plasma waste reforming plant for syngas production. The plant layout is based on the findings of chapter 6 and shown in fig. 7.3.

A rotary kiln combined with a plasma unit is the key component. A screw feeder with crusher feeds solids into the rotary kiln, ashes are caught in a designated storage. The pyrolysis gas that forms in the rotary kiln is cooled by a spray condenser to a temperature above the boiling point of water but low enough, so a solid state acid absorber can be integrated. The spray condenser also enriches the gas stream with steam and substitutes a boiler. It was verified in all operation points that this is possible: The heat energy in the gas stream is always higher than the necessary heat for vaporization of added water. A 2:1 stoichiometric ratio between solid waste and water is assumed. Organic gases that liquefy in the spray condenser are separated; they can be fed back into the rotary kiln. A fan circulates the steam enriched pyrolysis gas. A recriculation rate of 3 is assumed in all scenarios. The product gas undergoes an acid scrubber, which can be implemented as a second spray condenser, before a final base washing step to remove all halogen compounds. The syngas product stream is kept consistent at 100 TPD. The amount of halogenic compounds is assumed to be negligible in mass flow. Three parameters are used to describe the process: $X_{pyrolysis}$ describes the conversion of solids to pyrolysis gas. $X_{reforming}$ is the conversion efficiency at which hydrocarbons are reformed into syngas. The energy efficiency of the plasma reforming is $\eta_{reforming}$. All reactions are assumed to proceed stoichiometrically, which neglects the higher volatility of hydrogen compared to carbon. The amount of pyrolysis gas required to produce the set amount of syngas is

$$M_{pyrolysis-gas} = M_{syngas} \frac{1}{X_{reforming}} \frac{14\,g\,mol^{-1}}{32\,g\,mol^{-1}}. \qquad (7.2)$$

This amount of pyrolysis gas is proportional to the amount of gas that needs to be recirculated and thus affects component size and mass flow of water. All components with the exception of the plasma unit are affected by this. The screw feeder and crusher dimensions are further increased due to the conversion of the pyrolysis $X_{pyrolysis}$. This also affects the required mass of input material:

$$M_{input} = M_{pyrolysis-gas} \frac{1}{X_{pyrolysis}} \qquad (7.3)$$

The amount of ash generated is proportional to $1 - X_{pyrolysis}$, hence the storage also is scaled by this factor. All volume flows are calculated from mass flows using an average molar mass of $16\,g\,mol^{-1}$, which is the medium of products, feed and also close to the molar mass of water vapour and methane. All components of the plant are scaled in size for each scenario. Their cost was adapted from literature and is shown in table. 7.2.

Component	Reference capacity	Nominal cost in €	Installed base case	Installed cost in €
Rotary kiln [145]	200 TPD	3,200,000	50 TPD	985,000
Water pump [96]	65 TPD	3,000	65 TPD	3,000
Spray condenser [96]	65 TPD	12,000	65 TPD	12,000
Turbo blower [96]	8 NM^3/s	50,000	8 Nm^3/h	50,000
HCl solid absorber [96]	8 Nm^3/s	200,000	8 Nm^3/h	200,000
Acid scrubbing [145]	127 TPD	8,600	100 TPD	7,000
Base wash [145]	127 TPD	48,000	100 TPD	40,000
Screw feeder [145]	200 TPD	105,000	50 TPD	32,000
Crusher [145]	200 TPD	20,000	50 TPD	6,200
Char storage [145]	200 m^3	76,000	50 TPD	23,400
Plasma unit	3 kW	2,000	11.2 MW	2,180,000

Table 7.2: Cost of plant components found in literature along with capacity in standard cubic meters per second (NM^3/h), tons per day (TPD) cubic meter or kilowatt. Also given is a cost estimate for the proposed waste reforming plant in the scenario marked as 'optimistic' in table 7.3. The scaling method used is explained in the chapter 3. Component costs may vary throughout scenarios.

Optimizing the parameters $X_{reforming}$, $X_{pyrolysis}$ and $\eta_{reforming}$ is crucial in order to achieve an economically viable process. An inefficient power-to-X-process is bound to use too much electrical power. The parameters that can be achieved in a real world scenario are not yet known. However, a process evaluation can show how high the parameters must be to achieve economically viable operation - if it is possible at all. The highest conversion of methane to syngas by plasma achieved in literature was also measured in a gliding arc reactor by Meguernes et al. [146] at $X_{reforming} = 44\%$ at an energy efficiency of $\eta = 33\%$. These values are used as 'pessimistic' scenario. The conversion of solids to gas $X_{pyrolysis}$ is temperature dependent, in commercial pyrolysis plants it easily exceeds 90%, often reaching up to 97%. A value of $X_{pyrolysis} = 90\%$ is used as pessimistic scenario. All other scenarios use 95%. Improvements in energy efficiency and conversion of the reforming reaction by 50% are simulated to identify which parameter has greater influence on the resulting costs per ton of syngas. The scenario 'target' contains a slight further improvement in both. The scenario 'optimistic' contains an energy efficiency of $\eta_{reforming} = 60\%$ - when a higher efficiency is assumed, the waste heat is not necessarily sufficient to generate steam at a methane-to-water ratio of 3. All scenario parameters are shown in table 7.3. The resulting costs of syngas are illustrated by composition in fig. 7.4.

Case	$X_{reforming}$	$X_{pyrolysis}$	$\eta_{reforming}$	CapEx estimate	Installed power
Pessimistic	0.44	0.9	0.33	7.9 m €	23 MW
Improved conversion	0.66	0.95	0.33	6.9 m €	23 MW
Improved efficiency	0.44	0.95	0.5	6.6 m €	15 MW
Target	0.66	0.95	0.5	5.5 m €	14 MW
Optimistic	0.9	0.95	0.6	4.5 m €	13 MW

Table 7.3: Parameters used for the production cost evaluation of plasma waste reforming shown in fig. 7.4.

The costs for syngas in the scenario 'pessimistic' are very high at 603 € per ton. It appears that low conversion has a smaller impact on this fig.than low energy efficiency. Two positions are more affected by conversion: First, the CO_2 emissions are higher at low conversion, since unconverted hydrocarbons are attributed to CO_2-emissions when they would be burned. Second, the capital costs increase because the gas- and mass flows need to be large to make up for the low conversion.

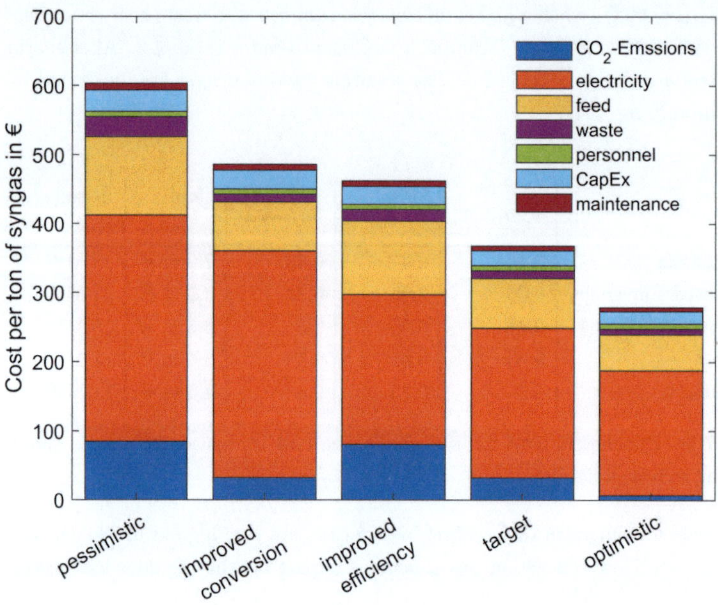

Figure 7.4: Calculated production cost composition of syngas by plasma waste reforming at varying plant performance using the process shown in figure 7.3.

The scenario 'target' reaches costs of 324 € per ton of syngas, while the optimistic scenario amounts to 278 € per ton. In all scenarios, electricity is responsible for the largest share of costs. Feedstock material and capital costs account for most of the remaining costs. The 'target' scenario is used in following comparisons to other technologies.

7.3 Other Processes

The conventional method for production of syngas is the steam reforming of methane (SRM). As such, it can't be omitted in this comparison. Data by Canete et al. [147] on the composition of operational cost and on capital cost is used in order to compare it to the presented processes in detail. The plant capital costs were scaled to arrive at a capacity of 100 TPD, while the plant analyzed by Canete et al. has a capacity of approximately 1,200 TPD. A scaled down plant is thus estimated to cost 4.11 million €. Personal expenses and maintenance expenses were added to these costs. CO_2 emissions were primarily derived from the conversion of the reformer, given that tail gas is burned to generate heat for the reformer. The conversion efficiency is 83.5%. In comparison to the 16.5 mol fraction of carbon that thus ends up as CO_2, the extra gas burned for heat can be neglected.

A similar approach is used on the co-electrolysis plant analysed in a paper by Quingxi et al. [148]. In a co-electrolysis, a feed stream composed of steam and CO_2 is applied to a high temperature electrolysis stack. The electrolyte is an oxygen ion conductor, the catalysts are picked so that both gases can be reduced to make syngas. The electrolysis stack is operated in isothermal mode, so that very little heat is utilized next to electrical energy. The analysed plant has a capacity of 5.6 TPD, the investment costs are adapted to a 100 TPD plant capacity. The scaled up co-electrolysis plant is thus estimated to cost 5.4 million €. Compared to the capital expenditures given by other sources for an alkaline electrolysis plant, this fig.seems low as it results in a similar cost per watt as previously estimated for the more established technology. Personnel expenses and maintenance expenses were added to these costs. The maintenance was increased to 10% of the capital costs per year to cover the short lifetime given for the electrolyzer stack of just 20,000 hours, or 2.5 years. Furthermore it has to be noted that no heat losses and perfect efficiency, except for losses in the driver, were assumed by Quingxi et al. [148]. The used feedstock for syngas production per ton were calculated from the given data for both processes. They are shown in table 7.4.

Resource per ton syngas	SMR	co-Electrolysis
CO_2 emissions	0.272 t	0.15 t
CO_2 feedstock	-	1.52 t
Natural gas	0.4875 t	-
Electricity	0.75 MWh	7.9 MWh
Water	1.9 t	1.2 t

Table 7.4: Resources used by steam reforming of methane (SMR) and co-electrolysis per ton of 2:1 syngas, adapted from [147, 148].

The required resources together with the capital expenditures and assumptions for
personnel expenditures and maintenance can be used to calculate the product cost.
The calculation is verified by repeating it at the scale and cost assumptions given
in literature and arriving at the same product cost given there. The composition of
product costs in the base case are illustrated in fig. 7.5.

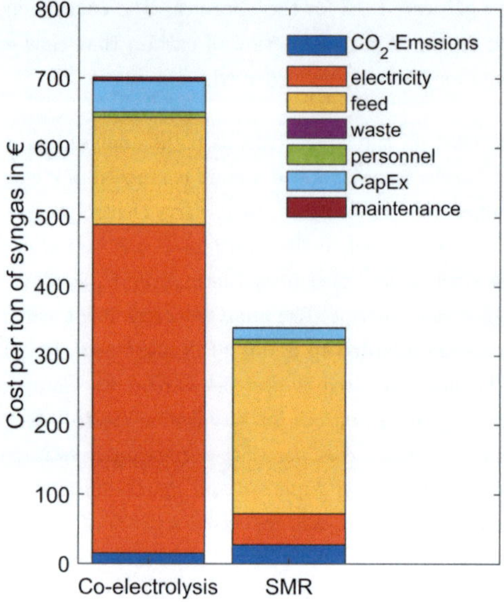

Figure 7.5: Calculated production cost composition of syngas by steam reforming
of methane (SMR) and co-electrolysis, in the base case, using the adjusted plant
costs and operational costs determined from literature in table 7.4.

The results show, that SMR has a vast cost advantage, syngas can be produced
for 340 € per ton compared to 701 € per ton for co-electrolysis. One main reason
is the yet low cost of CO_2 emission compared to the cost of carbon capture. Thus,
the feedstock for co-electrolysis actually accounts for similar cost compared to the
natural gas used as feedstock in SMR. The higher plant costs only have a minor
influence. Electricity accounts for the largest share of costs in co-electrolysis, while
only adding slightly to the cost of SMR.

7.4 Sensitivity Analysis

This section draws a comparison between the costs of syngas production of all discussed technologies. The sensitivity of these costs to the market price of feedstocks and energy is shown in fig. 7.6.

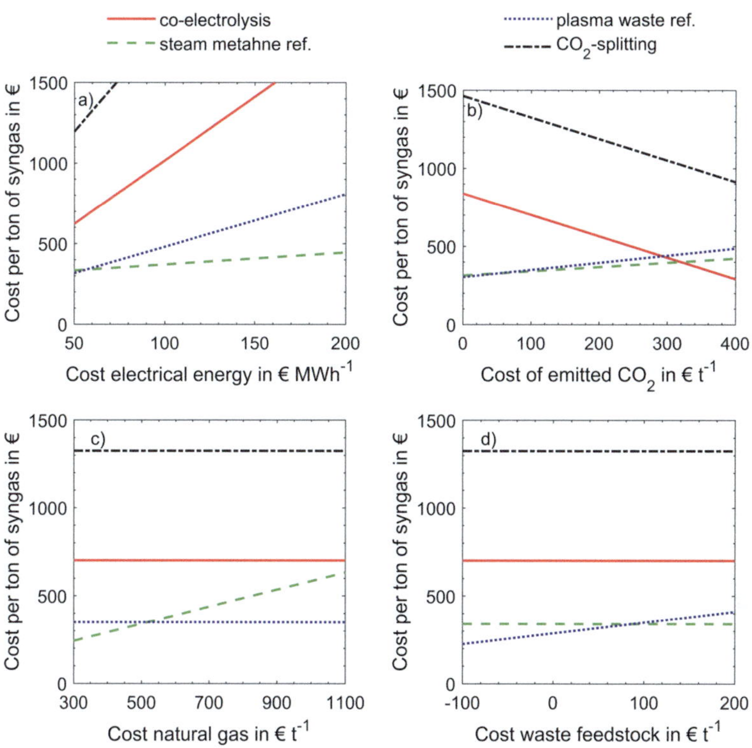

Figure 7.6: Sensitivity analysis of syngas production cost by various technologies. In a), the cost of electrical energy is varied. In b) CO_2 emission costs are varied, which also influences the cost of CO_2 as feedstock. c) shows variation in the price of natural gas, while in d) the cost of waste used as feedstock is varied.

Plasma waste reforming appears to reach the lowest levelized cost of production in the base scenario at 335 €t^{-1}. The established steam reforming of methane reaches the second lowest levelized cost of production in the base scenario at 340 € per ton. This value heavily depends on the cost of natural gas, which is shown in fig. 7.6 a). When natural gas is cheaper than 460 €t^{-1}, which corresponds to approximately 9.2 €GJ^{-1} or 33 € MWh^{-1}, SMR remains the cheapest process. Both, SMR and waste reforming are similarly affected by the costs associated to CO_2 emissions, which can be seen in b). SMR shows the lowest dependency on electrical energy cost, thus it remains cheaper when costs for electrical energy rise above 70 €MWh^{-1}, which is illustrated in c). If lower grade waste is utilized that can be obtained for free or even at a negative cost, the plasma reforming of waste becomes more economical. Conversely, if waste gains in price as recycling methods improve, SMR is cheaper at a price for waste material above 136 €t^{-1}, which is illustrated in d).

In comparison, co-electrolysis and plasma CO_2 splitting only become viable under extreme circumstances. Co-electrolysis is the cheapest technology when the costs associated with CO_2 emissions exceed 330 €t^{-1}. Under the assumptions adopted from [148], levelized costs of syngas made by co-electrolysis are consistently lower by a factor of two compared to a combination of CO_2 splitting and water electrolysis. However, these results must be taken with a grain of salt. When CO_2 splitting was excluded from the plant, just producing hydrogen via electrolysis still equated to a higher cost per volume of hydrogen than the cost calculated to make syngas via co-electrolysis. This extreme comparison casts some doubt on the true costs of co-electrolysis.

After comparing all results it must be concluded, that plasma waste reforming has the potential to be the cheapest alternative for syngas production under the premise that energy efficiency and conversion of the plasma steam reforming process can be advanced compared to literature values. When a 50% increase in both parameters can be achieved the process becomes viable under certain market conditions, i.e. rise in natural gas prices while electricity remains cheap. A further increase broadens the spectrum in which the process can be applied.

7.5 CO_2-Emissions

No comparison of sustainable technology is complete without a look on the carbon balance. This consideration must include indirect and direct carbon emissions. The major source of carbon emissions in electrical processes is of course the electrical energy itself. The specific emissions of electrical energy in Germany were $366\,kg\,MWh$ in 2022. Varying this value between 0 and $500\,kg\,MWh$ allows to illustrate the impact of electricity consumption on the emission balance, which is shown in fig. 7.7.

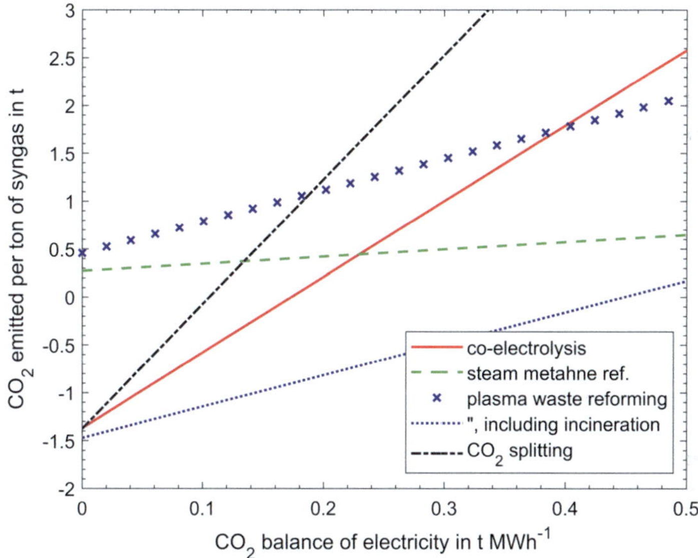

Figure 7.7: Sum of direct CO_2 emissions per ton syngas and emissions due to electrical energy consumption for different processes. Plasma waste reforming in shown twice, including and excluding the emissions that would otherwise arise from incineration.

It is apparent, that steam reforming of methane reaches the lowest emissions in this comparison given the current energy mix and the resulting specific emissions of electrical energy. When the specific CO_2 emissions are below $230\,kg\,MWh$, co-electrolysis becomes the more sustainable option. This makes sense, since it is in theory a CO_2-negative technology. The same is true for electrical splitting of CO_2. Due to imperfect conversion both capture less than the theoretical amount of CO_2 per ton. Plasma waste reforming is shown twice: The upper curve represents the carbon balance of the process, while the lower includes the fact that waste materials would otherwise be incinerated, which leads to more emissions. One might argue, that this is the relevant comparison - the produced syngas could well end up in new materials that inevitably become waste at some point. Thus, a circular process should consider the full life cycle of materials. Regardless, plasma waste reforming does lead to considerable emissions in this analysis. The main reason for this is the low conversion that was assumed: 66%, compared to over 80% in steam reforming of methane.

Of course, additional indirect emissions must be considered that are not included in this comparison: Natural gas suffers from a leakage rate during transportation that is not known exactly but estimated at up to 1% [149]. At the same time, methane has a global warming potential (GWP) of between 59 and 90. The GWP describes the impact of a greenhouse gas relative to CO_2. A 1% leakage rate and GWP for methane of 70 means, that each ton of methane that is utilized accounts for 0.18 tons of extra CO_2 equivalent [150]. Furthermore, natural gas often intrinsically contains CO_2 which is released during utilization - up to 30% can be found in some gas fields while the European Union only permits up to 1.5% [147, 151].

7.6 Summary

Methanol is produced from syngas in enormous amounts annually by steam reforming of natural gas. Making methanol is thus one of the major usages for carbon monoxide. Three alternative processes were analysed in this chapter to determine their cost and specific CO_2 emissions. To achieve this, process routes were analysed to gain insight into the capital expenditures necessary for each technology. The operation- and capital expenditures were calculated for varying performance parameters of the plasma reactors on a scale of 100 tons of syngas per day. This allows the derivation of 'target' parameters under which economic operation might become feasible. Lastly, the technologies were compared to each other in terms of cost and CO_2 emissions.

Plasma based CO_2 splitting combined with water electrolysis seems like a viable option for sustainable syngas production, since it is carbon negative. A process route was proposed in this chapter that uses a plasma process in combination with oxygen pressure swing absorption and an amine scrubber to produce carbon monoxide. Simulations show, that as of now the technology is too expensive to be applied. Furthermore, a high temperature co-electrolysis reaches lower cost, energy usage, and also lower specific CO_2 emissions. Both technologies are far from industrial scale and require immense amounts of electrical energy. The plasma reforming of waste material offers an alternative fossil-free approach. A process route was proposed that uses a rotary kiln in combination with a plasma unit, while steam is generated by using a spray evaporator. The consumption of electrical energy is lower by a factor of 4 compared to the combined approach of CO_2 splitting and electrolysis. Although the plant itself is more expensive than one for steam reforming, the feedstock is vastly cheaper. This results in cheaper operation costs when natural gas reaches prices in excess of 500 € per ton under the assumption that electrical energy is still available at costs below 70 € per MWh. For now, steam reforming achieves a higher conversion efficiency of hydrocarbons to syngas than plasma processes are able to for now. Thus, specific CO_2 emissions are higher in plasma waste reforming. However, there is no reason to expect that plasma processes will not be able to match thermal processes in the future in this regard. When the source of carbon - waste or fossil - is included in the comparison, plasma waste reforming offers the potential to be both cheaper and more environmentally friendly than the alternatives. This requires the target parameters to be reached: A plasmatron that can convert hydrocarbons to syngas at a conversion of 75% and energy efficiency of 55%.

Conclusion

The scope of this work was to tackle the problem of sustainable production of carbon monoxide by plasma for the synthesis of fuel and chemicals on multiple levels. Before industrial adaptation, plasma technology must find its way from theory and laboratory into engineering and large scale demonstration. The first chapters and the associated literature research therefore tried to illuminate the topic from as many sides as possible. Chemistry, physics and electrical engineering were the main focus. The insights gained were utilized in the construction and evaluation of two plasma sources. While dielectric barrier discharges were first pursued in the scope of this dissertation, they did not actually make it into the final thesis. The gliding arc reactor is better suited for the task. It has gained growing attention in the scientific community for application in plasma aided chemical processes. This is due to multiple reasons, foremost its ability to generate 'warm' plasma: Gas discharges with elevated gas temperature but non-thermal electron energy. One major issue is the ignition and upscaling of the technology. The first reactor that was constructed within the scope of this work allowed for a discharge power of 35 Watts, while the second allowed for up to 350 Watts. Both were constructed based on a similar concept: Direct current is applied to coaxial electrodes, a ring and a rod. A magnetic field in axial direction is applied, thus forcing a forming discharge channel to rotate. This could be observed to generate a large swept volume, i.e. the volume of the reactor in which gas is in contact to the discharge. The reactors are driven by a resonant inverter that can generate current limiting behaviour alongside a ferrite core transformer. To further increase the output voltage until a gas discharge can be ignited, a voltage multiplier cascade was used. In the second version of the driver, the multiplier features a dump load and is passively bypassed after ignition of the plasma. The driver circuits were developed along with the reactors and analyzed with them to gain insights into the interplay between electrical discharge and the driver circuit. The discharges that could be generated operate in the glow-to-arc transition region. In this operation mode, the differential electrical resistance of the discharge is negative - burn voltage decreases when current increases. The mode of operation was determined by measuring the distribution of electric field in the discharge. This data could also be used to determine power density of the plasma. Optical observations confirm that the discharge is made up of a rotating channel, although it appears as a disk to the naked eye. This channel does not lengthen considerably in response to the applied magnetic field. This is contrary to literature data and attributed to the driver circuit. In contrast to this work, most research uses voltage sources in series to ballast resistors. The drivers were able to operate the discharge stably in this region, although a repeating cycle of ignition and extinguishing of the discharge was found in some operation points. Most importantly, the energy efficiency of the driver could be determined to be up to 75% and ways to optimize it were suggested. This culminates in a new proposed design using an LLC converter to drive the plasma, which sadly could not be realized in time for this work. Furthermore the voltage/current characteristics could be determined in relation to the strength of the applied magnetic field. This data can be employed when designing scaled up systems in the future.

The constructed plasma reactors were used to split CO_2. The reactor is operated with gas flowing from top to bottom through the discharge plane, after which quenching occurs on a packed bed made from zirconium oxide balls. The approach proofed successful, reaching up to 45% energy efficiency at 28% conversion. The conversion seems to be mainly dependent on the applied power and only have a minor dependence on gas flow. The highest conversion was not achieved at the highest applied power. This could suggest that the conditions for splitting of CO_2 are ideal at this power, which might be the mean electron energy or gas temperature. The results are compared to literature results that used similar reactors and seem superior. Careful reproduction of the measurements and further studies are necessary to confirm them, since only rudimentary equipment was used in the experiments. The high efficiency in CO_2 splitting could be explained by a combination of several design choices that were made in the construction of the reactor which all proved successful in literature: Turbulent gas flow was avoided. Quenching on a ceramic material, that also serves as a catalyst, can increase yield and prevent back reaction. A co-linear electrode design was used. Magnets were used to guarantee for a large swept volume. None of the approaches were new, although they were contrived independently, but their combination is a new feat that was possible among other reasons due to the self-igniting driver circuit that was constructed. In spite of the satisfactory results that could be achieved, the splitting of CO_2 remains an energy intensive endeavour. A lot of waste heat is generated - heat that could be utilized. To achieve this, two processes were proposed that could utilize the high-temperature waste heat generated by a plasma. Both use the same gliding arc plasma technology that was previously proposed. The first one is the plasma based conversion of calcium carbonate into calcium oxide, carbon monoxide and oxygen. Calcium oxide releases carbon dioxide when heated, which is then split. The process relies on a circulation of the gas: While one part is extracted as a product, the majority is circulated through the plasma as a medium for heat transfer. An oxygen absorber can be integrated into the reactor to increase yield. Preliminary experiments are performed to test the viability, which show a qualitative success - more cannot be expected since a reactor was used that was originally devised for a gas phase process. Based on these experiments and the characteristics of the plasma reactor that were worked out before, a design for scaled up experiments was proposed and the basic design principles were discussed. The second process is the reforming of waste using plasma. In principle, the idea is not new. However, the gliding arc reactor design allows the integration of the plasma into the main reaction chamber. The plasma waste heat gasifies the solid material to organic gases. After recirculation, these gases are reformed in the plasma, while a part of the product gas is extracted. The idea is again tested in two preliminary experiments. The first is the dry reforming of methane, which lead to carbon deposition by methane pyrolysis as a significant side reaction. The second is the dry reforming of plastic waste, which also resulted in carbon formation but produced a clean product gas that almost exclusively consisted of carbon monoxide and dioxide. Both experiments showed, that carbon deposition at low temperature may pose a serious issue. Plasma aside, thermal equilibrium is expected to have a large influence on a reactor of this type. Two equilibria act on the working gas: The Boudouard reaction and the water gas shift. To avoid carbon deposition and obtain a high product gas stream, both are analysed for various feedstock composition and reactor temperature. This analysis along with the previously obtained characteristics of the gliding arc reactor can be used to design a scaled reactor that is suited for high efficiency plasma waste reforming.

Research alone cannot have any meaningful impact without implementation. For this reason, the plasma processes must also be viewed from an economical perspective. This was done on the example of syngas production for methanol synthesis. A simple plant was proposed based on plasma aided steam reforming of waste for the sake of analyzing the technology. Mass flows were calculated, which allowed the calculation of capital and operational expenditures and lastly production cost of syngas and associated emissions. The same was done for a combination of plasma CO_2 splitting and water electrolysis. The results were compared to literature data for steam reforming of methane and as well as co-electrolysis. A sensitivity analysis shows which technology achieves the lowest product costs under varying boundary conditions. In calculations, CO_2 splitting is consistently more expensive than co-electrolysis. This is in part due to the fact that an oxygen absorber and CO_2 recycling must be included into the plant for gas separation. The main reason, however, seems to be the assumed cost of water electrolysis. Steam reforming of natural gas is cheaper by a considerable margin than both processes. When electricity is available at moderate cost while natural gas remains expensive, plasma waste reforming was predicted to achieve lower costs while at the same time being a more sustainable technology.

These results underline a potential strength of plasma technology: No catalysts are used. This means that no rare metals are employed, but also that processes can be more robust and operate with impure feedstock. While electrolysis might reach higher energy efficiency in a lot of applications, this advantage enables new processes and ensures that plasma technology will have a place in the future of sustainable electrical chemistry.

Appendix A

Different Approaches to Conversion: A Short Proof

This appendix presents a short proof that the CO_2 conversion X calculated throughout this thesis is indeed correct and equivalent to the usually approach. It was originally written to satisfy a reviewer with attention for details. The usual formula for conversion in the CO_2 splitting equation is

$$X(A) = \frac{n_A(start) - n_A(end)}{n_A(start)} = \frac{n_{CO2}(start) - n_{CO2}(end)}{n_{CO2}(start)} . \qquad (A.1)$$

The amount of a substance n and concentration c are related by

$$n_i = c_i V \quad (1) \qquad (A.2)$$

with volume V, which is constant at either end of the reactor (however, $V(start)$ is not equal to $V(end)$, so n and c can be used interchangeably here. The amount of CO in the exhaust gas is equivalent to the amount in the supplied gas plus the amount of converted CO_2:

$$n_{CO}(end) = n_{CO}(start) + \Delta n \quad (2) \qquad (A.3)$$

When pure CO_2 is used as feed:

$$n_{CO}(end) = n_{CO2}(start) - n_{CO2}(end) \quad (3) \qquad (A.4)$$

To prove that

$$\frac{n_{CO2}(start) - n_{CO2}(end)}{n_{CO2}(start)} = \frac{c_{CO}(end)}{c_{CO}(end) + c_{CO2}(end)} \qquad (A.5)$$

148

we can use (1):

$$\frac{n_{CO2}(start) - n_{CO2}(end)}{n_{CO2}(start)} = \frac{n_{CO}(end)}{n_{CO}(end) + n_{CO2}(end)} \tag{A.6}$$

Using (3) yields

$$\frac{n_{CO2}(start) - n_{CO2}(end)}{n_{CO2}(start)} = \frac{n_{CO2}(start) - n_{CO2}(end)}{n_{CO}(end) + n_{CO2}(end)} \tag{A.7}$$

Using (3) again yields

$$\frac{n_{CO2}(start) - n_{CO2}(end)}{n_{CO2}(start)} = \frac{n_{CO2}(start) - n_{CO2}(end)}{n_{CO2}(start) - n_{CO2}(end) + n_{CO2}(end)}, \tag{A.8}$$

thus proving that both expressions are similar under the given assumptions.

Appendix B

Calculation of Conversion in Dry Reforming

This section lists the calculation performed to work out conversion of two reactions in the dry reforming experiment presented in chapter 6.2.

The two reactions that are considered are:

$$1)\, CH_4 + CO_2 \longrightarrow 2CO + 2H_2\,. \tag{B.1}$$

$$2)\, CH_4 \longrightarrow C + 2H_2\,. \tag{B.2}$$

Their conversions are X_1 and X_2 respectively. n_0 is the amount of CO_2 and CH_4, which are equal. The mole number of each compound measured downstream thus is:

$$n(CO) = 2\,X_1\,n_0 \;\rightarrow\; \frac{n(CO)}{n_0} = 2\,X_1 \tag{B.3}$$

$$n(H_2) = (2\,X_1\,2\,X_2)\,n_0 \;\rightarrow\; \frac{n(H_2)}{n_0} = 2\,X_1 + 2\,X_2 \tag{B.4}$$

$$n(CH_4) = n_0 - X_1\,n_0 - X_2\,n_0 \;\rightarrow\; \frac{n(CH_4)}{n_0} = 1 - X_1 - X_2 \tag{B.5}$$

$$n(CO_2) = n_0 - X_1\,n_0 \;\rightarrow\; \frac{n(CO_2)}{n_0} = 1 - X_1 \tag{B.6}$$

The sum of them is the mole number of products:

$$n_2 = n(CO) + n(H_2) + n(CH_4) + n(CO_2) = 2 + 2X_1 + X_2 \qquad \text{(B.7)}$$

The concentration of hydrogen and carbon monoxide are the ratios between $n(CO)$ respectively $n(H_2)$ and n_2:

$$c(H_2) = \frac{n(H_2)}{n_2} = \frac{2\,X_1 + 2\,X_2}{2 + 2\,X_1 + X_2}$$
$$\rightarrow\ c(H_2) + (c(H_2) - 1)\,X_1 + \left(\frac{c(H_2)}{2} - 1\right) X_2 = 0 \qquad \text{(B.8)}$$

$$c(CO) = \frac{n(CO)}{n_2} = \frac{2\,X_1}{2 + 2\,X_1 + X_2}$$
$$\rightarrow\ c(CO) + (c(CO) - 1)X_1 + \frac{c(CO)}{2}X_2 = 0 \qquad \text{(B.9)}$$

These equations can then be solved as an equation system with the measured concentrations of H_2 and CO. Solving yields X_1 and X_2. The concentration of CO_2 and CH_4 in the product gas can be calculated as well and compared to measurements. If they do not agree, additional reactions could account for the difference.

Appendix C

Plasma Source Cost Lineup

The following table contains a preliminary cost lineup of a $3000\,W$ plasma source using an LLC topology presented in this work. This is of course bound to change during further development of the technology. Note also that production cost, not sales price is given. This figure is meant to represent a lower bound of potential unit cost.

Item or service	Cost in €
Electrodes	70
Mantle	80
Mount	50
Ceramic foam diffusor	20
Thermal insulation	20
Insulating bushing	50
Plastic parts (mechanical)	20
Radiator	22
Heat pipes	15
Pipes and flanges, O-rings	25
Cooling fans	50
Washers and screws	10
Gas fittings	25
Case	100
PCBs	10
PCB assembly	10
HV-tranformer	50
HV-rectifier	30
Magnetic components	50
ICs & Other components	80
Sensors	15
Controller & regulation	30
20% Overhead for mechanical assembly	106
20% Overhead for testing	106
System integration, proportionately	250
Relais and fuses	100
PFC-rectifier, proportionately at 0.2 €/kW	600
Sum	1929

Appendix D

Preliminary Results of the LLC Plasma Driver

This chapter gives a quick overview of preliminary results that show the feasibility of an LLC converter as a plasma source. A converter was designed and built alongside a custom transformer. The design resonance Frequency was $f_r = 100\,kHz$, the transformer ratio is $1:50$. The plasma source is thus able to provide up to $10\,kV$ output voltage and up to $600\,mA$ output current. The equations presented in section 2.3.4 were used to simulate and tune the behaviour of the driver circuit. The gain G at varying frequency is shown in fig. D.1.

The simulated voltage-current characteristics are shown in fig. D.2. It can be seen that the short circuit current of the output reaches a value of less than $600\,mA$ at a frequency of $F_x = 1.6$ in the simulation. A variable resistor $(0...10\,k\Omega)$ can be used to measure the characteristics of the driver circuit.

The measured gain at varying quality factor can be seen in fig. D.3. The general trend is similar to the expected values. However, peak at $F_x = 1$ is less pronounced than expected. This is not an issue since it lies out of the design operation frequency range.

The measured voltage-current characteristics can be seen in fig. D.4. The measured current is slightly lower than predicted. This is due to the MOSFET deadtimes that reduce the effective dutycycle, which were not considered in the simulation. However, current limiting can be seen to work very good: The V/I curves are almost vertical. Sadly, no higher resistances were available at the time, so no full V/I-curve could be obtained.

The preliminary results show the feasibility of the proposed LLC-converter as a

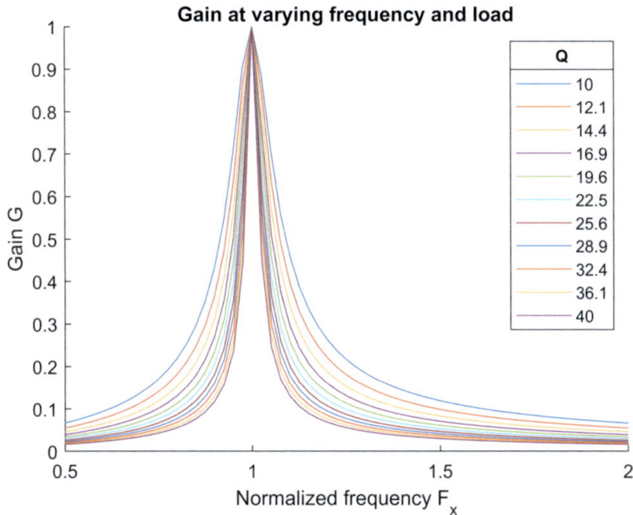

Figure D.1: Simulated normalized voltage gain of the LLC converter at varying normalized frequency F_x and quality factor Q.

driver circuit for a gliding arc plasma.

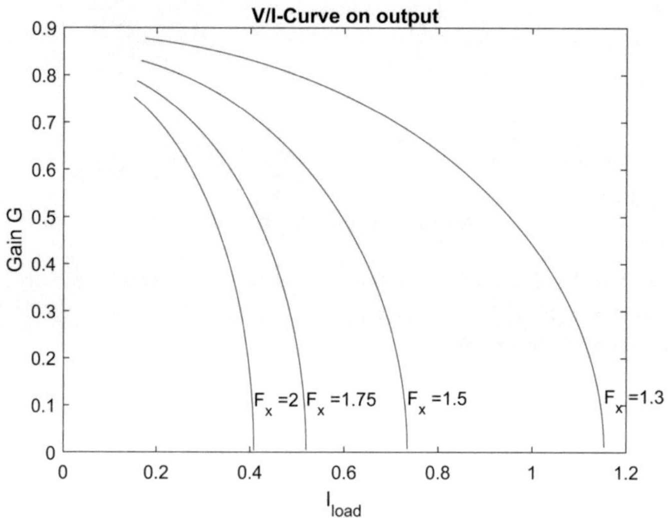

Figure D.2: Simulated normalized voltage gain of the LLC converter at varying normalized frequency F_x and load current I_{load} in A.

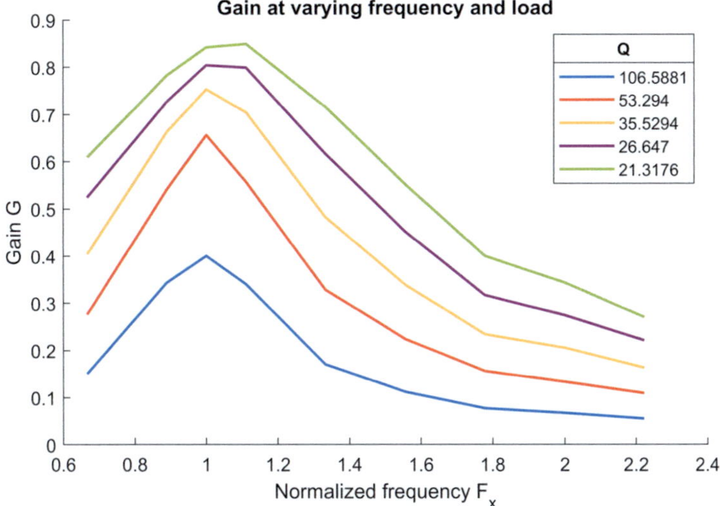

Figure D.3: Measured normalized voltage gain of the LLC converter at varying normalized frequency F_x and quality factor Q.

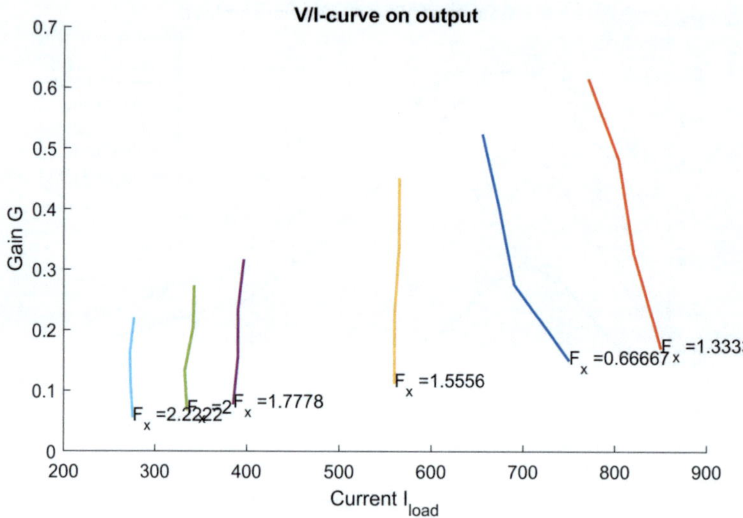

Figure D.4: Measured normalized voltage gain of the LLC converter at varying normalized frequency F_x and load current I_{load} in mA.

Abbreviations

AC	Alternating current
AFOLU	Agriculture, Forrest and Land Use
C1	Molecules with one carbon atom
C2	Molecules with two carbon atoms
CapEx	Capital Expenditures
CCS	Carbon Capture and Storage
CDR	Carbon Dioxide Removal
D	Diode (in circuit diagrams)
DBD	Dielectric Barrier Discharge
DC	Direct current
ExB	physics jargon: Electric and mangetic field perpendicular
GA	Gliding Arc
GC	Gas Chromatographer
HV	High Voltage
IPCC	Intergovernmental Panel on Climate Change
kV	Kilo volts

LC	Resonance circuit with one inductance and one capacitance
LLC	Resonance circuit with two inductances and one capacitance
M	MOSFET (in circuit diagrams)
mk	mark, used to identify versions of a product
MW	Microwave
NDIR	Non-Dispersive Infra Red
OpEx	Operational Expenditures
PSA	Pressure Swing Absorber
RF	Radio frequency
RWGS	Reverse Water Gas Shift
SCC	Standard Cubic Centimeter
SCCM	Standard Cubic Centimeters per Minute
SLM	Standard Liter per Minute
SRM	Steam Reforming of Methane
TPD	Tons Per Day)
TVS	Transient Voltage Suppressor (in circuit diagrams)
UV	Ultraviolet

Symbols

α	Degree of ionization	-
λ	Mean free path	m
λ	Thermal conductivity	$\mathrm{W\,m^{2-1}\,K^{-1}}$
μ_e	Electron mobility	$\mathrm{m^2\,V^{-1}\,s^{-1}}$

ρ	Density	$\mathrm{kg\,m^{-3}}$
σ	Conductivity	$\Omega\,m$
c_p	Heat capacity at constant pressure	$\mathrm{J\,mol^{-1}\,K^{-1}}$
D	Diameter	m
d	Distance, specifically discharge gap	m
J	Current density	$\mathrm{J\,m^{-2}}$
j_+	Current density of ions	$\mathrm{J\,m^{-2}}$
j_i	Current density of electrons	$\mathrm{J\,m^{-2}}$
n	Particle density	$\mathrm{m^{-3}}$
n_+	Positive particle density	$\mathrm{m^{-3}}$
n_-	Negative particle density	$\mathrm{m^{-3}}$
n_i	Ionized particle density	$\mathrm{m^{-3}}$
n_n	Neutral particle density	$\mathrm{m^{-3}}$
p	Pressure	Pa
p_{amb}	Ambient pressure	Pa
Q	Heat flux	W
R	Resistance	Ω
R_{sup}	Resistance of supply	Ω
V_M	Molar volume	$\mathrm{SCC\,mol^{-1}}$
$\dot{M}$	Mass flow	TPD, tons per day
$\dot{V}$	Flow rate	SCCM, SLM
ϵ	Strain of the discharge channel	m
η	Energy efficiency	-

Γ	Component cost	Euro
$\hat{i}$	Current ripple	A
$\hat{V}$	Peak voltage	V
ω	angular velocity	s^{-1}
A	Area	m^2
a	Annuity factor	-
a	Turn ratio of transformer	-
B	Magnetic field	T
c	Concentration	-
c_d	drag coefficient	-
c_i	Concentration of substance i	-
C_r	Resonance capacitance	F
E	Electric field	$\mathrm{V\,m}^{-1}$
E_B	Breakdown electric field	$\mathrm{V\,m}^{-1}$
F	Force	N
f	Frequency	s^{-1}
F_D	Drag force	N
F_L	Lorentz force	N
f_r	Resonance frequency	s^{-1}
g	Transconductance	-
G_r	Resonance gain	-
H	Enthalpy	$\mathrm{kJ\,mol}^{-1}$
H^0	Enthalpy at standard conditions	$\mathrm{kJ\,mol}^{-1}$

H_f	Enthalpy of formation	$\mathrm{kJ\,mol^{-1}}$
H_r	Enthalpy of reaction	$\mathrm{kJ\,mol^{-1}}$
H_{vol}	Enthalpy per volume	$\mathrm{kJ\,mol^{-1}}$
I	Current	A
$I_{discharge}$	Discharge current	A
k	Reaction rate constant	A.U.
k_B	Boudouard reaction rate constant	A.U.
$k_{steam-ref}$	Steam reforming reaction rate constant	A.U.
k_{WGS}	Water gas shift reaction rate constant	A.U.
L	Inductance	H
L_m	Magnetization inductance	H
L_p	Primary side inductance	H
L_r	Resonance inductance	H
L_s	Secondary side inductance	H
M	Mass	kg
M_e	Electron rest mass	kg
N	Turn number of transformer	-
N_{stages}	Number of stages in multiplier circuit	-
P	Power	W
P_{vol}	Volumetric power density	$\mathrm{W\,m^{-3}}$
Q	Quality factor	-
R	Recirculation ratio	-
r_G	Gyration radius	m

r_i	Ionic radius	m
R_p	Primary side resistance	Ω
R_s	Secondary side resistance	Ω
S	Selectivity	-
T	Temperature	K
t	time	s
T_e	Electron temperature	eV
T_p	Period	W
T_{rot}	Period of rotation	W
V	Voltage	V
v	Speed of reaction	A.U.
v	Velocity	T
V_B	Breakdown voltage	V
v_e	Electron velocity	T
V_{burn}	Burn voltage, i.e. voltage between electrodes	A
$V_{measured}$	Measured voltage	V
W	Energy, work	J
X	Conversion	-
Y	Yield	-
Z	Impedance	-

Constants

$1\,Td$	Townsend (unit)	$10 \times 10^{-21}\ \mathrm{V\,m^{-1}}$

k_B	Boltzmann constant	1.380×10^{-23} J K^{-1}
n_0	Ideal number density at standard pressure	2.7×10^{-25} m^{-3}
N_A	Avogadros number	6.022×10^{23} mol^{-1}
q, q_e	Elementary charge	1.602×10^{-19} A s
q, q_e	Faraday Constant	96485 A s mol^{-1}
R	Ideal gas constant	8.314 J K^{-1} mol^{-1}

List of Tables

List of Figures

Bibliography

[1] I. E. Agency, *Energy technology transitions for industry, p. 30*, 2009.

[2] intergovernmental panel on climate change, "Mitigation of climate change, chapter 8", Contribution of working group III to the fifth assessment report of the intergovernmental panel on climate change **1454**, 147 (2014).

[3] I. E. Agency, *Energy technology transitions for industry, p. 105*, 2009.

[4] M. Allen, P. Antwi-Agyei, F. Aragon-Durand, M. Babiker, P. Bertoldi, M. Bind, S. Brown, M. Buckeridge, I. Camilloni, A. Cartwright, et al., "Technical summary: global warming of 1.5° c. an ipcc special report on the impacts of global warming of 1.5° c above pre-industrial levels and related global greenhouse gas emission pathways, in the context of strengthening the global response to the threat of climate change, sustainable development, and efforts to eradicate poverty", (2019).

[5] intergovernmental panel on climate change, "Mitigation of climate change, chapter 2.3.2", Contribution of working group III to the fifth assessment report of the intergovernmental panel on climate change **1454**, 147 (2014).

[6] A. Mtibe, M. P. Motloung, J. Bandyopadhyay, and S. S. Ray, "Synthetic biopolymers and their composites: advantages and limitations—an overview", Macromolecular Rapid Communications **42**, 2100130 (2021).

[7] O. Adeyeye, E. R. Sadiku, A. Babu Reddy, A. S. Ndamase, G. Makgatho, P. S. Sellamuthu, A. B. Perumal, R. B. Nambiar, V. O. Fasiku, I. D. Ibrahim, et al., "The use of biopolymers in food packaging", in *Green biopolymers and their nanocomposites* (Springer, 2019), pp. 137–158.

[8] D. Garraın, R. Vidal, P. Martınez, V. Franco, R. González, et al., "How'green'are biopolymers?", in Ds 42: proceedings of iced 2007, the 16th international conference on engineering design, paris, france, 28.-31.07. 2007 (2007), pp. 229–230.

[9] X. Wu and G. Chen, "Global overview of crude oil use: from source to sink through inter-regional trade", Energy Policy **128**, 476–486 (2019).

[10] M. Parsaee, M. K. D. Kiani, and K. Karimi, "A review of biogas production from sugarcane vinasse", Biomass and bioenergy **122**, 117–125 (2019).

[11] A. Antelava, N. Jablonska, A. Constantinou, G. Manos, S. A. Salaudeen, A. Dutta, and S. M. Al-Salem, "Energy potential of plastic waste valorization: a short comparative assessment of pyrolysis versus gasification", Energy & Fuels **35**, 3558–3571 (2021).

[12] J. Fu, Y. Huang, Q. Liao, A. Xia, Q. Fu, and X. Zhu, "Photo-bioreactor design for microalgae: a review from the aspect of co2 transfer and conversion", Bioresource Technology **292**, 121947 (2019).

[13] E. Klemm, C. M. Lobo, A. Löwe, V. Schallhart, S. Renninger, L. Waltersmann, R. Costa, A. Schulz, R.-U. Dietrich, L. Möltner, et al., "Chemampere: technologies for sustainable chemical production with renewable electricity and CO_2, N_2, O_2, and H_2O", The Canadian Journal of Chemical Engineering.

[14] R. Snoeckx and A. Bogaerts, "Plasma technology–a novel solution for CO_2 conversion?", Chemical Society Reviews **46**, 5805–5863 (2017).

[15] C.-j. Liu, R. Mallinson, and L. Lobban, "Nonoxidative methane conversion to acetylene over zeolite in a low temperature plasma", Journal of catalysis **179**, 326–334 (1998).

[16] C. Mesters, "A selection of recent advances in c1 chemistry", Annual Review of Chemical and Biomolecular Engineering **7**, 223–238 (2016).

[17] Tom from Ex&F, *Remember benzene? it's back!*, (2022) `https://youtu.be/I-ZawsbYpmo?t=304` (visited on 03/16/2022).

[18] Q. Meng, J. Yan, R. Wu, H. Liu, Y. Sun, N. Wu, J. Xiang, L. Zheng, J. Zhang, and B. Han, "Sustainable production of benzene from lignin", Nature Communications **12**, 1–12 (2021).

[19] T. Brinkmann, G. G. Santonja, F. Schorcht, S. Roudier, and L. D. Sancho, "Best available techniques (bat) reference document for the production of chlor-alkali", Publ. Off. Eur. Union (2014).

[20] R. I. Masel, Z. Liu, H. Yang, J. J. Kaczur, D. Carrillo, S. Ren, D. Salvatore, and C. P. Berlinguette, "An industrial perspective on catalysts for low-temperature CO_2 electrolysis", Nature Nanotechnology **16**, 118–128 (2021).

[21] K. Li, J.-L. Liu, X.-S. Li, X.-B. Zhu, and A.-M. Zhu, "Post-plasma catalytic oxidative CO_2 reforming of methane over ni-based catalysts", Catalysis Today **256**, 96–101 (2015).

[22] A. Minet, "The electric furnace: its origin, transformations and applications. part ii", Transactions of the Faraday Society **1**, 85–102 (1905).

[23] E. J. Houston, "On some early forms of electric furnaces", Journal of the Franklin Institute **125**, 475–479 (1888).

[24] E. V. Hort and P. Taylor, "Acetylene-derived chemicals", Kirk-Othmer Encyclopedia of Chemical Technology (2000).

[25] W. E. Hanford and D. L. Fuller, "Acetylene chemistry", Industrial & Engineering Chemistry **40**, 1171–1177 (1948).

[26] H. Schobert, "Production of acetylene and acetylene-based chemicals from coal", Chemical reviews **114**, 1743–1760 (2014).

[27] H. S. Eyde, "The manufacture of nitrates from the atmosphere by the electric arc—birkeland-eyde process", Journal of the Royal Society of Arts **57**, 568–576 (1909).

[28] M. B. Rubin, "The history of ozone. the schönbein period, 1839–1868", Bull. Hist. Chem **26**, 40–56 (2001).

[29] A. D. Miller, W. R. Grow, L. A. Dees, and T. J. Manning, "A history of patented methods of ozone production from 1897 to 1997", (2002).

[30] M. Biondi, "Microwave gas discharges", Electrical Engineering **69**, 806–809 (1950).

[31] A. Fridman and V. Rusanov, "Theoretical basis of non-equilibrium near atmospheric pressure plasma chemistry", Pure and applied chemistry **66**, 1267–1278 (1994).

[32] J. Loreiro and J. Amorim, "Kinetics and spectroscopy of low temperature plasma", in *Kinetics and spectroscopy of low temperature plasmas* (Springer, 2016), pp. 3–10.

[33] J. Loreiro and J. Amorim, "Kinetics and spectroscopy of low temperature plasma", in *Kinetics and spectroscopy of low temperature plasmas* (Springer, 2016), pp. 10–17.

[34] S. Pancheshnyi, M. Nudnova, and A. Starikovskii, "Development of a cathode-directed streamer discharge in air at different pressures: experiment and comparison with direct numerical simulation", Physical Review E **71**, 016407 (2005).

[35] P. K. Chu and X. Lu, *Low temperature plasma technology: methods and applications* (CRC press, 2013), pp. 5–6.

[36] J. Loreiro and J. Amorim, "Kinetics and spectroscopy of low temperature plasma", in *Kinetics and spectroscopy of low temperature plasmas* (Springer, 2016), pp. 16–40.

[37] P. K. Chu and X. Lu, *Low temperature plasma technology: methods and applications* (CRC press, 2013), pp. 19–24.

[38] P. K. Chu and X. Lu, *Low temperature plasma technology: methods and applications* (CRC press, 2013), pp. 41–97.

[39] W. Witteman, "Rotational-vibrational structure of co 2", in *The co2 laser* (Springer, 1987), pp. 8–52.

[40] W. D. Callister et al., *Fundamentals of materials science and engineering*, Vol. 471660817 (Wiley London, 2000), pp. 723–733.

[41] M. W. Chase Jr, "Nist-janaf thermochemical tables", J. Phys. Chem. Ref. Data, Monograph **9** (1998).

[42] L. T. Petkovska, "Sequence and hot transitions in the co2 molecule; interference of adjacent lines belonging to different transitions as a function of temperature", Spectroscopy Letters **33**, 283–300 (2000).

[43] A. Bogaerts, T. Kozák, K. Van Laer, and R. Snoeckx, "Plasma-based conversion of co 2: current status and future challenges", Faraday discussions **183**, 217–232 (2015).

[44] L. B. Loeb, *Electrical coronas, their basic physical mechanisms* (Univ of California Press, 1965), pp. 2–3.

[45] L. B. Loeb, *Electrical coronas, their basic physical mechanisms* (Univ of California Press, 1965), pp. 4–12.

[46] L. B. Loeb, *Electrical coronas, their basic physical mechanisms* (Univ of California Press, 1965), pp. 22–26.

[47] R. H. Fowler and L. Nordheim, "Electron emission in intense electric fields", Proceedings of the Royal Society of London. Series A, Containing Papers of a Mathematical and Physical Character **119**, 173–181 (1928).

[48] P. K. Chu and X. Lu, *Low temperature plasma technology: methods and applications* (CRC press, 2013), pp. 25–41.

[49] W. Gambling and H. Edels, "The high-pressure glow discharge in air", British Journal of Applied Physics **5**, 36 (1954).

[50] L. B. Loeb, *Electrical coronas, their basic physical mechanisms* (Univ of California Press, 1965), pp. 22–26.

[51] L. B. Loeb, *Fundamental processes of electrical discharge in gases* (J. Wiley & Sons, Incorporated, 1939).

[52] T. Rokunohe, Y. Yagihashi, F. Endo, and T. Oomori, "Fundamental insulation characteristics of air; n2, co2, n2/o2, and sf6/n2 mixed gases", Electrical Engineering in Japan **155**, 9–17 (2006).

[53] S. Kumar, T. Huiskamp, A. Pemen, M. Seeger, J. Pachin, and C. M. Franck, "Electrical breakdown study in CO_2 and CO_2-O_2 mixtures in AC, DC and pulsed electric fields at 0.1–1 MPa pressure", IEEE Transactions on Dielectrics and Electrical Insulation **28**, 158–166 (2021).

[54] Y. Kabouzi, M. Calzada, M. Moisan, K. Tran, and C. Trassy, "Radial contraction of microwave-sustained plasma columns at atmospheric pressure", Journal of applied physics **91**, 1008–1019 (2002).

[55] M. Shneider, M. Mokrov, and G. Milikh, "Dynamic contraction of the positive column of a self-sustained glow discharge in molecular gas", Physics of Plasmas **19**, 033512 (2012).

[56] U. Stroth, *Plasmaphysik* (Springer, 2011), pp. 15–32.

[57] U. Stroth, *Plasmaphysik* (Springer, 2011), pp. 261–299.

[58] A. Fridman, A. Gutsol, S. Gangoli, Y. Ju, and T. Ombrello, "Characteristics of gliding arc and its application in combustion enhancement", Journal of Propulsion and Power **24**, 1216–1228 (2008).

[59] U. Stroth, *Plasmaphysik* (Springer, 2011), pp. 317–334.

[60] T. Kozák and A. Bogaerts, "Splitting of CO_2 by vibrational excitation in non-equilibrium plasmas: a reaction kinetics model", Plasma Sources Science and Technology **23**, 045004 (2014).

[61] A. Bogaerts and G. Centi, "Plasma technology for CO_2 conversion: a personal perspective on prospects and gaps", Frontiers in Energy Research **8**, 111 (2020).

[62] A. Hofmann, *Physical chemistry essentials* (Springer, 2018), pp. 167–182.

[63] K. G. Denbigh and K. G. Denbigh, *The principles of chemical equilibrium: with applications in chemistry and chemical engineering* (Cambridge University Press, 1981), pp. 40–43.

[64] R. Snoeckx and A. Bogaerts, "Plasma technology–a novel solution for CO_2 conversion?", Chemical Society Reviews **46**, 5805–5863 (2017).

[65] G. Trenchev and A. Bogaerts, "Dual-vortex plasmatron: a novel plasma source for co2 conversion", Journal of CO2 Utilization **39**, 101152 (2020).

[66] H. Kim, S. Song, C. P. Tom, and F. Xie, "Carbon dioxide conversion in an atmospheric pressure microwave plasma reactor: improving efficiencies by enhancing afterglow quenching", Journal of CO2 Utilization **37**, 240–247 (2020).

[67] B. Raja, R. Sarathi, and R. Vinu, "Development of a swirl-induced rotating glow discharge reactor for co2 conversion: fluid dynamics and discharge dynamics studies", Energy Technology **8**, 2000535 (2020).

[68] S. Andreev, V. Zakharov, V. Ochkin, and S. Y. Savinov, "Plasma-chemical co2 decomposition in a non-self-sustained discharge with a controlled electronic component of plasma", Spectrochimica Acta Part A: Molecular and Biomolecular Spectroscopy **60**, 3361–3369 (2004).

[69] S. C. Kim, M. S. Lim, and Y. N. Chun, "Reduction characteristics of carbon dioxide using a plasmatron", Plasma Chemistry and Plasma Processing **34**, 125–143 (2014).

[70] W. Bongers, H. Bouwmeester, B. Wolf, F. Peeters, S. Welzel, D. van den Bekerom, N. den Harder, A. Goede, M. Graswinckel, P. W. Groen, et al., "Plasma-driven dissociation of co2 for fuel synthesis", Plasma processes and polymers **14**, 1600126 (2017).

[71] L. F. Spencer and A. D. Gallimore, "Efficiency of co2 dissociation in a radio-frequency discharge", Plasma Chemistry and Plasma Processing **31**, 79–89 (2011).

[72] A. Ozkan, T. Dufour, T. Silva, N. Britun, R. Snyders, F. Reniers, and A. Bogaerts, "Dbd in burst mode: solution for more efficient co2 conversion?", Plasma Sources Science and Technology **25**, 055005 (2016).

[73] K. G. Denbigh and K. G. Denbigh, *The principles of chemical equilibrium: with applications in chemistry and chemical engineering* (Cambridge University Press, 1981), pp. 133–146.

[74] R. Patel, *Experimental studies on lignite gasification process* (Ahmedabad Nirma Institute of Technology, 2015).

[75] C. A. Callaghan, *Kinetics and catalysis of the water-gas-shift reaction: a microkinetic and graph theoretic approach* (Worcester Polytechnic Institute, 2006).

[76] G. Franchi, M. Capocelli, M. De Falco, V. Piemonte, and D. Barba, "Hydrogen production via steam reforming: a critical analysis of mr and rmm technologies", Membranes **10**, 10 (2020).

[77] S. N. Makarov, R. Ludwig, and S. J. Bitar, *Practical electrical engineering* (Springer, 2019), pp. 52–60.

[78] Z. Vukic, *Nonlinear control systems* (CRC Press, 2003), pp. 50–55.

[79] D. R. Young, "Electric breakdown in co2 from low pressures to the liquid state", Journal of Applied Physics **21**, 222–231 (1950).

[80] D. Drosihn, "Übersicht zur entwicklung der energiebedingten emissionen und brennstoffeinsätze in deutschland 1990–2018", Dessau-Roßlau: Herausgeber Umweltbundesamt (2020).

[81] M. Willert-Porada, *Advances in microwave and radio frequency processing* (Springer, 2006), pp. 85–91.

[82] Y. Wang, W. Ding, J. Yan, and Y. Wang, "A diffusive atmospheric pressure glow discharge in a coaxial pin-to-ring gap with a transverse magnetic field", AIP Advances **7**, 095209 (2017).

[83] P. Bruggeman, J. Liu, J. Degroote, M. G. Kong, J. Vierendeels, and C. Leys, "Dc excited glow discharges in atmospheric pressure air in pin-to-water electrode systems", Journal of Physics D: Applied Physics **41**, 215201 (2008).

[84] R. H. Stark and K. H. Schoenbach, "Direct current glow discharges in atmospheric air", Applied Physics Letters **74**, 3770–3772 (1999).

[85] J. B. Dance, *Cold cathode tubes* (Iliffe, 1968), pp. 25–27.

[86] G. H. Royer and R. L. Bright, *Patent for an electrical inverter circuits*, US Patent 2,849,614, Aug. 1958.

[87] P. Baxandall, "Transistor sine-wave lc oscillators. some general considerations and new developments", Proceedings of the IEE-Part B: Electronic and Communication Engineering **106**, 748–758 (1959).

[88] I. Barbi and F. Pöttker, *Soft commutation isolated dc-dc converters*, Vol. 1 (Springer, 2019), pp. 141–186.

[89] B. Yang, F. C. Lee, and M. Concannon, "Over current protection methods for llc resonant converter", in Eighteenth annual ieee applied power electronics conference and exposition, 2003. apec'03. Vol. 2 (IEEE, 2003), pp. 605–609.

[90] S. Abdel-Rahman, "Resonant llc converter: operation and design", Infineon Technologies North America (IFNA) Corp **19** (2012).

[91] H.-S. Choi, "Design consideration of half-bridge llc resonant converter", Journal of power electronics **7**, 13–20 (2007).

[92] T. Tanzawa, *On-chip high-voltage generator design* (Springer, 2013), pp. 17–65.

[93] M. J. Johnson, D. R. Boris, T. B. Petrova, and S. G. Walton, "Extending the volume of atmospheric pressure plasma jets through the use of additional helium gas streams", Plasma Sources Science and Technology **29**, 015006 (2020).

[94] S. Renninger, M. Lambarth, and K. P. Birke, "High efficiency co2-splitting in atmospheric pressure glow discharge", Journal of CO2 Utilization **42**, 101322 (2020).

[95] S. Renninger, J. Stein, M. Lambarth, and K. P. Birke, "An optimized reactor for co2 splitting in dc atmospheric pressure discharge", Journal of CO2 Utilization **58**, 101919 (2022).

[96] M. S. e. a. Peters, *Eex co2 prices, auction market*, (2022) http://www.mhhe.com/engcs/chemical/peters/data/ (visited on 03/16/2022).

[97] EEX-AG, *Eex natural gas price price reference egix*, (2022) `https://www.eex.com/en/markets/natural-gas` (visited on 03/16/2022).

[98] EEX-AG, *Eex co2 prices, auction market*, (2022) `https://www.eex.com/en/market-data/environmental-markets/auction-market` (visited on 03/16/2022).

[99] T. Reinhardt and U. Richers, *Entsorgung von schredderrückständen-ein aktueller überblick* (Forschungszentrum Karlsruhe, 2004).

[100] H. S. Kwak, H. S. Uhm, Y. C. Hong, and E. H. Choi, "Disintegration of carbon dioxide molecules in a microwave plasma torch", Scientific reports **5**, 1–13 (2015).

[101] S. Watanabe, S. Saito, K. Takahashi, and T. Onzawa, "Observations of the glow-to-arc transition between thoriated tungsten electrodes with a parallel rc circuit", Journal of Physics D: Applied Physics **36**, 2521 (2003).

[102] N. Dyatko, Y. Z. Ionikh, I. Kochetov, D. Marinov, A. Meshchanov, A. Napartovich, F. Petrov, and S. Starostin, "Experimental and theoretical study of the transition between diffuse and contracted forms of the glow discharge in argon", Journal of Physics D: Applied Physics **41**, 055204 (2008).

[103] W. Xia, L. Li, Y. Zhao, Q. Ma, B. Du, Q. Chen, and L. Cheng, "Dynamics of large-scale magnetically rotating arc plasmas", Applied physics letters **88**, 211501 (2006).

[104] C. Wang, H. Cui, Z. Zhang, W. Xia, and W. Xia, "Observation of arc modes in a magnetically rotating arc plasma generator", Contributions to Plasma Physics **57**, 395–403 (2017).

[105] H. Zhang, L. Li, X. Li, W. Wang, J. Yan, and X. Tu, "Warm plasma activation of CO_2 in a rotating gliding arc discharge reactor", Journal of CO2 Utilization **27**, 472–479 (2018).

[106] C. Wang, Z. Lu, D. Li, W. Xia, and W. Xia, "Effect of the magnetic field on the magnetically stabilized gliding arc discharge and its application in the preparation of carbon black nanoparticles", Plasma Chemistry and Plasma Processing **38**, 1223–1238 (2018).

[107] S. Gangoli, A. Gutsol, and A. Fridman, "A non-equilibrium plasma source: magnetically stabilized gliding arc discharge: i. design and diagnostics", Plasma Sources Science and Technology **19**, 065003 (2010).

[108] S. Sun, H. Wang, D. Mei, X. Tu, and A. Bogaerts, "CO_2 conversion in a gliding arc plasma: performance improvement based on chemical reaction modeling", Journal of CO_2 Utilization **17**, 220–234 (2017).

[109] Z. Liu, H. Yang, R. Kutz, and R. I. Masel, "CO_2 electrolysis to CO and O_2 at high selectivity, stability and efficiency using sustainion membranes", Journal of The Electrochemical Society **165**, J3371 (2018).

[110] H. Zhang, L. Li, R. Xu, J. Huang, N. Wang, X. Li, and X. Tu, "Plasma-enhanced catalytic activation of CO_2 in a modified gliding arc reactor", Waste Disposal & Sustainable Energy **2**, 139–150 (2020).

[111] M. Ramakers, J. A. Medrano, G. Trenchev, F. Gallucci, and A. Bogaerts, "Revealing the arc dynamics in a gliding arc plasmatron: a better insight to improve CO_2 conversion", Plasma Sources Science and Technology **26**, 125002 (2017).

[112] L. Li, H. Zhang, X. Li, X. Kong, R. Xu, K. Tay, and X. Tu, "Plasma-assisted CO_2 conversion in a gliding arc discharge: improving performance by optimizing the reactor design", Journal of CO_2 Utilization **29**, 296–303 (2019).

[113] Y. Uytdenhouwen, K. Bal, I. Michielsen, E. Neyts, V. Meynen, P. Cool, and A. Bogaerts, "How process parameters and packing materials tune chemical equilibrium and kinetics in plasma-based CO_2 conversion", Chemical Engineering Journal **372**, 1253–1264 (2019).

[114] I. Michielsen, Y. Uytdenhouwen, J. Pype, B. Michielsen, J. Mertens, F. Reniers, V. Meynen, and A. Bogaerts, "CO_2 dissociation in a packed bed dbd reactor: first steps towards a better understanding of plasma catalysis", Chemical Engineering Journal **326**, 477–488 (2017).

[115] L. Li, H. Zhang, X. Li, J. Huang, X. Kong, R. Xu, and X. Tu, "Magnetically enhanced gliding arc discharge for CO_2 activation", Journal of CO_2 Utilization **35**, 28–37 (2020).

[116] M. Bai, Z. Zhang, X. Bai, M. Bai, and W. Ning, "Plasma synthesis of ammonia with a microgap dielectric barrier discharge at ambient pressure", IEEE transactions on plasma science **31**, 1285–1291 (2003).

[117] J. Hong, S. Prawer, and A. B. Murphy, "Production of ammonia by heterogeneous catalysis in a packed-bed dielectric-barrier discharge: influence of argon addition and voltage", IEEE Transactions on Plasma Science **42**, 2338–2339 (2014).

[118] M. D. Farahani, Y. Zeng, and Y. Zheng, "The application of nonthermal plasma in methanol synthesis via CO_2 hydrogenation", Energy Science & Engineering (2022).

[119] H. Lee and D. H. Kim, "Direct methanol synthesis from methane in a plasma-catalyst hybrid system at low temperature using metal oxide-coated glass beads", Scientific Reports **8**, 1–8 (2018).

[120] Q. Zhang, G. Zhang, L. Wang, X. Wang, S. Wang, and Y. Chen, "Measurement of the electron density in a microwave plasma torch at atmospheric pressure", Applied Physics Letters **95**, 201502 (2009).

[121] N. Chalyavi, P. S. Doidge, R. J. Morrison, and G. B. Partridge, "Fundamental studies of an atmospheric-pressure microwave plasma sustained in nitrogen for atomic emission spectrometry", Journal of Analytical Atomic Spectrometry **32**, 1988–2002 (2017).

[122] M. Kuchenbecker, N. Bibinov, A. Kaemlimg, D. Wandke, P. Awakowicz, and W. Viöl, "Characterization of dbd plasma source for biomedical applications", Journal of Physics D: Applied Physics **42**, 045212 (2009).

[123] J.-L. Liu, X. Wang, X.-S. Li, B. Likozar, and A.-M. Zhu, "CO_2 conversion, utilisation and valorisation in gliding arc plasma reactors", Journal of Physics D: Applied Physics **53**, 253001 (2020).

[124] N. Kovačić, G. A. Meyer, L. Ke-Ling, and R. M. Barnes, "Diagnostics in an air inductively coupled plasma", Spectrochimica Acta Part B: Atomic Spectroscopy **40**, 943–957 (1985).

[125] S. K. Vyas, R. K. Verma, S. Maurya, and V. Singh, "Review of magnetron developments", Frequenz **70**, 455–462 (2016).

[126] P. Jain, M. Recchia, M. Cavenago, U. Fantz, E. Gaio, W. Kraus, A. Maistrello, and P. Veltri, "Evaluation of power transfer efficiency for a high power inductively coupled radio-frequency hydrogen ion source", Plasma Physics and Controlled Fusion **60**, 045007 (2018).

[127] S. Renninger, P. Rößner, J. Stein, M. Lambarth, and K. P. Birke, "Towards high efficiency CO_2 utilization by glow discharge plasma", Processes **9**, 2063 (2021).

[128] Q. Huang, D. Zhang, D. Wang, K. Liu, and A. W. Kleyn, "Carbon dioxide dissociation in non-thermal radiofrequency and microwave plasma", Journal of Physics D: Applied Physics **50**, 294001 (2017).

[129] G. Kaur, A. P. Kulkarni, and S. Giddey, "CO_2 reduction in a solid oxide electrolysis cell with a ceramic composite cathode: effect of load and thermal cycling", International Journal of Hydrogen Energy **43**, 21769–21776 (2018).

[130] P. C. Zonetti, S. Letichevsky, A. B. Gaspar, E. F. Sousa-Aguiar, and L. G. Appel, "The $Ni_xCe_{075}Zr_{0.25-x}O_2$ solid solution and the rwgs", Applied Catalysis A: General **475**, 48–54 (2014).

[131] S. Li, M. Ongis, G. Manzolini, and F. Gallucci, "Non-thermal plasma-assisted capture and conversion of CO_2", Chemical Engineering Journal **410**, 128335 (2021).

[132] G. Giammaria, G. Van Rooij, and L. Lefferts, "Plasma catalysis: distinguishing between thermal and chemical effects", Catalysts **9**, 185 (2019).

[133] O. S. AB, *Stainless steel handbook* (Outokumpu Oy, 2013), pp. 56–60.

[134] S. Nema and K. Ganeshprasad, "Plasma pyrolysis of medical waste", Current science, 271–278 (2002).

[135] L. Tang, H. Huang, H. Hao, and K. Zhao, "Development of plasma pyrolysis/gasification systems for energy efficient and environmentally sound waste disposal", Journal of Electrostatics **71**, 839–847 (2013).

[136] H. Huang and L. Tang, "Treatment of organic waste using thermal plasma pyrolysis technology", Energy Conversion and Management **48**, 1331–1337 (2007).

[137] P. G. Rutberg, "Plasma pyrolysis of toxic waste", Plasma physics and controlled fusion **45**, 957 (2003).

[138] N. Miskolczi, L. Bartha, and G. Deák, "Thermal degradation of polyethylene and polystyrene from the packaging industry over different catalysts into fuel-like feed stocks", Polymer degradation and stability **91**, 517–526 (2006).

[139] C. Guéret, M. Daroux, and F. Billaud, "Methane pyrolysis: thermodynamics", Chemical Engineering Science **52**, 815–827 (1997).

[140] trading economics, *Eu natural gas front-month futures historical price data*, (2022) `https://tradingeconomics.com/commodity/eu-natural-gas` (visited on 03/16/2022).

[141] T. Nguyen, Z. Abdin, T. Holm, and W. Mérida, "Grid-connected hydrogen production via large-scale water electrolysis", Energy conversion and management **200**, 112108 (2019).

[142] I. Heininger, "Vergleich und modellierung eines geeigneten elektrolyseverfahrens für einen power-to-x-prozess", (2021).

[143] W. Castle, "Air separation and liquefaction: recent developments and prospects for the beginning of the new millennium", International Journal of Refrigeration **25**, 158–172 (2002).

[144] J. Husebye, A. L. Brunsvold, S. Roussanaly, and X. Zhang, "Techno economic evaluation of amine based CO_2 capture: impact of CO_2 concentration and steam supply", Energy Procedia **23**, 381–390 (2012).

[145] V. Chhabra, A. Parashar, Y. Shastri, and S. Bhattacharya, "Techno-economic and life cycle assessment of pyrolysis of unsegregated urban municipal solid waste in india", Industrial & Engineering Chemistry Research **60**, 1473–1482 (2021).

[146] K. Meguernes, J. Chapelle, and A. Czernichowski, "Valorization of methane in electric arcs and high pressure cold discharges", High Temperature Material Processes: An International Quarterly of High-Technology Plasma Processes **5** (2001).

[147] B. Cañete, C. E. Gigola, and N. B. Brignole, "Synthesis gas processes for methanol production via ch4 reforming with co2, h2o, and o2", Industrial & Engineering Chemistry Research **53**, 7103–7112 (2014).

[148] Q. Fu, C. Mabilat, M. Zahid, A. Brisse, and L. Gautier, "Syngas production via high-temperature steam/co 2 co-electrolysis: an economic assessment", Energy & Environmental Science **3**, 1382–1397 (2010).

[149] V. Quaschning, *Specific carbon dioxide emissions of various fuels*, (2022) `https://www.volker-quaschning.de/datserv/CO2-spez/index_e.php` (visited on 03/16/2022).

[150] J. Lynch, M. Cain, R. Pierrehumbert, and M. Allen, "Demonstrating gwp*: a means of reporting warming-equivalent emissions that captures the contrasting impacts of short-and long-lived climate pollutants", Environmental Research Letters **15**, 044023 (2020).

[151] M. van Stiphout, *Gas quality standards in the european union, published by the international gas union*, (2022) `http://members.igu.org/html/wgc2009/papers/docs/wgcFinal00785.pdf` (visited on 03/16/2022).

Publications

Papers

1. Renninger, S., Lambarth, M. and Birke, K.P., 2020. High efficiency CO_2-splitting in atmospheric pressure glow discharge. Journal of CO_2 Utilization, 42, p.101322.

2. Renninger, S., Rößner, P., Stein, J., Lambarth, M. and Birke, K.P., 2021. Towards High Efficiency CO_2 Utilization by Glow Discharge Plasma. Processes, 9(11), p.2063.

3. Renninger, S., Stein, J., Lambarth, M. and Birke, K.P., 2022. An optimized reactor for CO_2 splitting in DC atmospheric pressure discharge. Journal of CO_2 Utilization, 58, p.101919.

4. Klemm, E., Lobo, C.M., Löwe, A., Schallhart, V., Renninger, S., Waltersmann, L., Costa, R., Schulz, A., Dietrich, R.U., Möltner, L. and Meynen, V., CHEMampere: Technologies for sustainable chemical production with renewable electricity and CO_2, N_2, O_2, and H_2O. The Canadian Journal of Chemical Engineering.

Conferences

1. 'pathway to a CO_2-neutral mobility: Plasma Fuel', Plenary Presentation, KEROGREEN Workshop for Plasma catalysis for renewable Fuels and Chemicals, 2019.

2. 'Plasma based CO_2 Utilization – A Comparison', SICT 2022 / Plasma Tech 2022 joint virtual session, in collaboration with Paul Rößner.

Poster

1. Renninger, S., Lambarth, M. and Birke, K.P., 2020. CO_2-Reduktion durch Plasma mit elektrochemischer Gastrennung zur nachhaltigen Herstellung von Kohlenstoffmonoxid. Chemie Ingenieur Technik, 92(9), pp.1335-1335.

Inventions

1. Schaltungsanordnung für eine Plasmaquelle zur Erzeugung von Plasma bei Atmosphärendruck, 2022

2. Feststoff-Plasma-Reaktor und Verfahren zum Betreiben eines Festbettreaktors, 2022

3. Verfahren und Vorrichtung zur Gewinnung von Kohlenmonoxid aus atmosphärischem Kohlendioxid mit einem Feststoff-Plasma-Reaktor, 2022

Students thesis and projects

1. 'Design eines Teststandes zur Untersuchung von Plasma durch dielektrische Barrierenentladung', Bachelorarbeit, Fabian Ney, 2019.

2. Entwicklung einer Sauerstoffpumpe für einen CO_2-Spaltungsprozess, Bachelorarbeit, Antonia Maria Mauch, 2020.

3. Entwicklung und Aufbau einer Spannungsquelle zur CO_2-Spaltung durch dielektrische Barrierenentladung, Masterarbeit, Johannes Gilsdorf, 2020

4. Entwicklung einer Systemversorgung und Steuerung für eine Gasdiffusionselektrode, Bachelorarbeit, Mahmoud Mardini, 2020

5. Aufbau eines durchsichtigen DBE-Plasmareaktors zur Spaltung von Kohlensotffdioxid, Projektarbeit, Daniel Draxler; Achim Johannes; Johannes Kretzschmar; Felix Maurer; Alicia Mörsdorf, 2020.

6. Wasserstoffbatterie mit oxidischen Interkalationsanoden, Yvonne Eboumbou Ebongue, 202

7. Modellierung und techno-ökonomische Bewertung einer plasmainduzierten Herstellung von Kraftstoff aus Überschussstrom und Prozessabgasen, Masterarbeit, Samuel Kaufmann, 2021.

8. Vergleich und Modellierung eines geeigneten Elektrolyseverfahrens für einen Power-to-X-Prozess, Masterarbeit, Iris Heiniger, 2021.

9. Optimierung einer Palsmaquelle zur CO_2-Spaltung in Barriereentladung, Bachelorarbeit, Felix Pander, 2021.

10. Optimierung von CO_2-Spaltung durch Glimmentladungsplasma, Bachelorarbeit, Jan Samuel Stein, 2021.

Curriculum Vitae

Persönliche Angaben

Name	Stephan Renninger
Geburtsdatum	22. April 1994
Geburtsort	Leonberg

Studium und Promotion

10/2018 - 04/2022	Wissenschaftlicher Mitarbeiter am Institut für Photovoltaik, Lehrstuhl für elektrische Energiespeichersysteme, Universität Stuttgart
10/2016 - 09/2018	Master of Science, Nachhaltige Elektrische Energieversorgung, Universität Stuttgart
10/2012 - 09/2015	Bachelor of Science, Erneuerbare Energien, Universität Stuttgart

Beruflicher Werdegang

seit 05/2022	Ausgründudgn im Rahmen des Exist-Forschungsfransfer-Programms, Cyclize, ipv, Universität Stuttgart
2018	Wissenscahftliche Hilfskraft, Entwicklung von Feststoffbatterien, ipv Universität Stuttgart
2017 - 2018	Werksstudent, Simulation und Softwareentwicklung für stationäre Batteriespeicher, Robert Bosch GmbH
2016 - 2017	Werksstudent, Fachplanung Gebäudetechnik, Krebs Ingenieure GmbH

Erklärung

Ich versichere hiermit, dass ich die vorliegende Arbeit selbstständig verfasst und keine anderen als die angegebenen Hilfsmittel verwendet habe.

Stuttgart, February 19, 2023

Stephan Renninger

Acknowledgements

I would like to thank my supervisor Peter Birke, who has the ability to constantly challenge his employees with new tasks, but also gives them the time they need when it really matters. He took the time for weekly meetings, even back when I was still working on my master thesis, and never hesitated to discuss out-of-the-box ideas (his and mine). He was a dependable and quick decider in strategic questions throughout my work at the institute. My thanks also goes to Very Meynen, who was willing to take the time to co-report on this thesis. Since I previously had the honor of co-authoring a paper with her, I know her to be thorough and precise, and I am very happy she was willing to take the time. I am grateful to my predecessors and colleagues at the institute for photovoltaics, who paved the way for this work by acquiring funding and sharing space in the laboratory. The value of my group of colleagues who are always ready to take the load of each others backs in tough times cannot be overstated. Special thanks goes to my colleagues who worked with me on plasma-related projects: Maike Lambarth, who was there from the beginning, Paul Rößner, who we managed to convince to join us as a group leader after befriending him at the ITC, and our former students Jan Stein and Samuel Kaufmann. I think we as a team managed to pull off a lot with very little.

The last years at the institute were stressful and often involved working late and not answering the phone, so I also would like to thank my family, especially my parents, Marion and Volker, and my sister Hannah, for being understanding when I called back late (or forgot to do so). My gratitude also goes to my proof-readers who probably saved me quite some embarrassment: Michael Rehme, Julia Pross-Brakhage, Felix Krebs, Jan Stein, Maike Lambarth and Samuel Kaufmann. Honorable mention: Timo Kropp, who is the archetype of the type of nerd that I strive to be, and who was happy to discuss ideas with me on video calls for years after leaving the institute. Furthermore I would like to thank the students who worked on projects with me in the institute. Working with them was a motivation to keep expanding my horizon in order to always have enough of a leading edge to be able to answer their questions.

The pandemic forced us all into home office, which made it hard to ask for advise when needed. Luckily, Michael Rehme was always there to help with his advise as a postdoc surrogate, roommate, band colleague and mountaineering companion. Lastly I would like to express my gratitude again to Maike Lambarth, who I cannot appreciate enough for her roles both within and outside of the institute. Without her this thesis would have been a lot less successful and probably would have taken a year longer.